First Course on Fuzzy Theory and Applications

T0181889

Advances in Soft Computing

Editor-in-chief
Prof. Janusz Kacprzyk
Systems Research Institute
Polish Academy of Sciences
ul. Newelska 6
01-447 Warsaw, Poland
E-mail: kacprzyk@ibspan.waw.pl

Further books of this series can be found on our homepage: springeronline.com

Rainer Hampel, Michael Wagenknecht,
Nasredin Chaker (Eds.)
Fuzzy Control
2000. ISBN 3-7908-1327-3

Henrik Larsen, Janusz Kacprzyk,
Sławomir Zadrozny, Troels Andreasen,
Henning Christiansen (Eds.)
Flexible Query Answering Systems
2000. ISBN 3-7908-1347-8

Robert John and Ralph Birkenhead (Eds.)
Developments in Soft Computing
2001. ISBN 3-7908-1361-3

Mieczysław Kłopotek, Maciej Michalewicz
and Sławomir T. Wierzchoń (Eds.)
Intelligent Information Systems 2001
2001. ISBN 3-7908-1407-5

Antonio Di Nola and Giangiacomo Gerla (Eds.)
Lectures on Soft Computing and Fuzzy Logic
2001. ISBN 3-7908-1396-6

Tadeusz Trzaskalik and Jerzy Michnik (Eds.)
Multiple Objective and Goal Programming
2002. ISBN 3-7908-1409-1

James J. Buckley and Esfandiar Eslami
An Introduction to Fuzzy Logic and Fuzzy Sets
2002. ISBN 3-7908-1447-4

Ajith Abraham and Mario Köppen (Eds.)
Hybrid Information Systems«
2002. ISBN 3-7908-1480-6

Przemysław Grzegorzewski, Olgierd Hryniewicz,
María ç. Gil (Eds.)
*Soft Methods in Probability, Statistics
and Data Analysis*
2002. ISBN 3-7908-1526-8

Lech Polkowski
Rough Sets
2002. ISBN 3-7908-1510-1

Mieczysław Kłopotek, Maciej Michalewicz
and Sławomir T. Wierzchoń (Eds.)
Intelligent Information Systems 2002
2002. ISBN 3-7908-1509-8

Andrea Bonarini, Francesco Masulli
and Gabriella Pasi (Eds.)
Soft Computing Applications
2002. ISBN 3-7908-1544-6

Leszek Rutkowski, Janusz Kacprzyk (Eds.)
Neural Networks and Soft Computing
2003. ISBN 3-7908-0005-8

Jürgen Franke, Gholamreza Nakhaeizadeh,
Ingrid Renz (Eds.)
Text Mining
2003. ISBN 3-7908-0041-4

Tetsuzo Tanino, Tamaki Tanaka,
Masahiro Inuiguchi
*Multi-Objective Programming and Goal
Programming*
2003. ISBN 3-540-00653-2

Mieczysław Kłopotek, Sławomir T. Wierzchoń,
Krzysztof Trojanowski (Eds.)
Intelligent Information Processing and Web Mining
2003. ISBN 3-540-00843-8

Ahmad Lotfi, Jonathan M. Garibaldi (Eds.)
Applications and Science in Soft-Computing
2004. ISBN 3-540-40856-8

Mieczysław Kłopotek, Sławomir T. Wierzchoń,
Krzysztof Trojanowski (Eds.)
*Intelligent Information Processing and
Web Mining*
2004. ISBN 3-540-21331-7

Miguel López-Díaz, María ç. Gil, Przemysław
Grzegorzewski, Olgierd Hryniewicz, Jonathan
Lawry
*Soft Methodology and Random Information
Systems*
2004. ISBN 3-540-22264-2

Kwang H. Lee
First Course on Fuzzy Theory and Applications
2005. ISBN 3-540-22988-4

Kwang H. Lee

First Course
on Fuzzy Theory
and Applications

With 149 Figures

 Springer

Dr. Kwang H. Lee
Advanced Institute of Science and Technology, KAIST
Kusong-dong 373-1
305-701 Taejon
Republic of South Korea

Library of Congress Control Number: 2004112297

ISSN 16-15-3871

ISBN 3-540-22988-4 Springer Berlin Heidelberg NewYork

Springer is a part of Springer Science+Business Media

springeronline.com

© Springer-Verlag Berlin Heidelberg 2005

Printed in Germany

Cover design: Erich Kirchner, Heidelberg
Typesetting: Digital data supplied by the authors

Printed on acid-free paper   62/3020Rw-5 4 3 2 1 0

Preface

Fuzzy theory has become a subject that generates much interest among the courses for graduate students. However, it was not easy to find a suitable textbook to use in the introductory course and to recommend to the students who want to self-study. The main purpose of this book is just to meet that need.

The author has given lectures on the fuzzy theory and its applications for ten years and continuously developed lecture notes on the subject. This book is a publication of the modification and summary of the lecture notes.

The fundamental idea of the book is to provide basic and concrete concepts of the fuzzy theory and its applications, and thus the author focused on easy illustrations of the basic concepts. There are numerous examples and figures to help readers to understand and also added exercises at the end of each chapter.

This book consists of two parts: a theory part and an application part. The first part (theory part) includes chapters from 1 to 8. Chapters 1 and 2 introduce basic concepts of fuzzy sets and operations, and Chapters 3 and 4 deal with the multi-dimensional fuzzy sets. Chapters 5 and 6 are extensions of the fuzzy theory to the number and function, and Chapters 7 and 8 are developments of fuzzy properties on the probability and logic theories.

The second part is for applications. Chapter 9 introduces fuzzy inference techniques which can be used in uncertain situations, and Chapter 10 is for the application of the inference to the control problems and expert systems. Chapters 11 and 12 provide possible hybrid combinations with other intelligent algorithms, especially neural network and genetic algorithms.

Special acknowledgements are due to my students who gave me suggestions and feedback on the lecture notes. I am also indebted to a series of grants from the Korea Foundation of Science and Technology, the Mirae Company, and the CHUNG Moon Soul BioInformation and BioElectronics Center.

Kwang H. LEE

KAIST (Korea Advanced Institute of Science and Technology)

Table of Contents

Chapter 1. FUZZY SETS

This chapter reviews the concepts and notations of sets (crisp sets), and then introduces the concepts of fuzzy sets. The concept of fuzzy sets is a generalization of the crisp sets. Convex set, α - cut operation, cardinality of fuzzy set and fuzzy number are also introduced.

1.1 Sets

1.1.1 Elements of Sets

An universal set X is defined in the universe of discourse and it includes all possible elements related with the given problem. If we define a set A in the universal set X, we see the following relationships

$$A \subseteq X.$$

In this case, we say a set A is included in the universal set X. If A is not included in X, this relationship is represented as follows.

$$A \nsubseteq X.$$

If an element x is included in the set A, this element is called as a member of the set and the following notation is used.

$$x \in A.$$

If the element x is not included in the set A, we use the following notation.

$$x \notin A.$$

In general, we represent a set by enumerating its elements. For example, elements $a_1, a_2, a_3, \ldots, a_n$ are the elements of set A, it is represented as follows.

$$A = \{a_1, a_2, \ldots, a_n\}.$$

Another representing method of sets is given by specifying the conditions of elements. For example, if the elements of set B should satisfy the conditions $P_1, P_2, \ldots, P_n$, then the set B is defined by the following.

$$B = \{b \mid b \text{ satisfies } p_1, p_2, \ldots, p_n\}.$$

In this case the symbol "$\mid$" implies the meaning of "such that".

In order to represent the size of N-dimension Euclidean set, the number of elements is used and this number is called cardinality. The cardinality of set A is denoted by $/A/$. If the cardinality $/A/$ is a finite number, the set A is a finite set. If $/A/$ is infinite, A is an infinite set. In general, all the points in N-dimensional Euclidean vector space are the elements of the universal set X.

1.1.2 Relation between Sets

A set consists of sets is called a *family of sets*. For example, a family set containing sets $A_1, A_2, \ldots$ is represented by

$$\{A_i / i \in I\}$$

where i is a set identifier and I is an identification set. If all the elements in set A are also elements of set B, A is a subset of B.

$$A \subseteq B \quad \text{iff (if and only if)} \quad x \in A \Rightarrow x \in B.$$

The symbol $\Rightarrow$ means "implication". If the following relation is satisfied,

$$A \subseteq B \quad \text{and} \quad B \subseteq A$$

A and B have the same elements and thus they are the same sets. This relation is denoted by

$$A = B$$

If the following relations are satisfied between two sets A and B,

$$A \subseteq B \quad \text{and} \quad A \neq B$$

then B has elements which is not involved in A. In this case, A is called a *proper subset* of B and   this relation is denoted by

$$A \subset B$$

A set that has no element is called an *empty set* $\varnothing$. An empty set can be a subset of any set.

1.1.3 Membership

If we use membership function (characteristic function or discrimination function), we can represent whether an element x is involved in a set A or not.

Definition (Membership function) For a set A, we define a membership function μ_A such as

$$\mu_A(x) = 1 \quad \text{if and only if} \quad x \in A$$
$$0 \quad \text{if and only if} \quad x \notin A.$$

We can say that the function μ_A maps the elements in the universal set X to the set $\{0,1\}$.

$$\mu_A : X \rightarrow \{0,1\}. \quad \square$$

As we know, the number of elements in a set A is denoted by the cardinality $|A|$. A power set $P(A)$ is a family set containing the subsets of set A. Therefore the number of elements in the power set $P(A)$ is represented by

$$|P(A)| = 2^{|A|}.$$

Example 1.1 If $A = \{a, b, c\}$, then $|A| = 3$

$$P(A) = \{\varnothing, \{a\}, \{b\}, \{a, b\}, \{a, c\}, \{b, c\}, \{a, b, c\}\}$$
$$|P(A)| = 2^3 = 8. \quad \square$$

1.2 Operation of Sets

1.2.1 Complement

The *relative complement set* of set A to set B consists of the elements which are in B but not in A. The complement set can be defined by the following formula.

$$B - A = \{x \mid x \in B, x \notin A\}.$$

If the set B is the universal set X, then this kind of complement is an absolute complement set $\overline{A}$. That is, $\overline{A} = X - A$

In general, a complement set means the absolute complement set. The complement set is always involutive

$$\overline{\overline{A}} = A.$$

The complement of an empty set is the universal set, and vice versa.

$$\overline{\varnothing} = X$$
$$\overline{X} = \varnothing.$$

1.2.2 Union

The union of sets A and B is defined by the collection of whole elements of A and B.

$$A \cup B = \{x \mid x \in A \text{ or } x \in B\}.$$

The union might be defined among multiple sets. For example, the union of the sets in the following family can be defined as follows.

$$\bigcup_{i \in I} A = \{x \mid x \in A_i \text{ for some } i \in I\} \text{ where the family of sets is } \{A_i / i \in I\}.$$

The union of certain set A and universal set X is reduced to the universal set.

$$A \cup X = X$$

The union of certain set A and empty set $\varnothing$ is A.
$$A \cup \varnothing = A$$
The union of set A and its complement set is the universal set
$$A \cup \overline{A} = X.$$

1.2.3 Intersection

The intersection $A \cap B$ consists of whose elements are commonly included in both sets A and B.
$$A \cap B = \{x \mid x \in A \text{ and } x \in B\}.$$
The Intersection can be generalized between the sets in a family of sets

$$\bigcap_{i \in I} A_i = \{x \mid x \in A_i, \quad \forall i \in I\} \text{ where } \{A_i / i \in I\} \text{ is a family of sets.}$$

The intersection between set A and universal set X is A.
$$A \cap X = A.$$
The intersection of A and empty set is empty set
$$A \cap \varnothing = \varnothing.$$
The intersection of A and its complement is all the time empty set
$$A \cap \overline{A} = \varnothing.$$
When two sets A and B have nothing in common, the relation is called as disjoint. Namely, it is when the intersection of A and B is empty set
$$A \cap B = \varnothing.$$

1.2.4 Partition of Set

Definition (Partition) A decomposition of set A into disjoint subsets whose union builds the set A is referred to a partition. Suppose a partition of A is π,
$$\pi(A) = \{A_i / i \in I, A_i \subseteq A\}$$
then A_i satisfies following three conditions.

i) $A_i \neq \varnothing$

ii) $A_i \cap A_j = \varnothing, \quad i \neq j, \quad i, j \in I$

iii) $\bigcup_{i \in I} A_i = A$ □

If there is no condition of (2), $\pi(A)$ becomes a cover or covering of the set A.

1.3 Characteristics of Crisp Set

1.3.1 Ordinary Characteristics

Let us look over the operational characteristics of union, intersection, and complement set [Table 1.1]. ***Commutativity*** of union and intersection is satisfied as follows

$$A \cup B = B \cup A$$
$$A \cap B = B \cap A.$$

The operations of intersection and union follows the ***associativity***

$$A \cup B \cup C = (A \cup B) \cup C = A \cup (B \cup C)$$
$$A \cap B \cap C = (A \cap B) \cap C = A \cap (B \cap C).$$

Union or intersection between itselves is reduced to the set itself. This is '***idempotency***'.

$$A \cup A = A$$
$$A \cap A = A.$$

In addition, for union and intersection, the distributivity is held.

$$A \cap (B \cup C) = (A \cap B) \cup (A \cap C)$$
$$A \cup (B \cap C) = (A \cup B) \cap (A \cup C).$$

De Morgan's law is satisfied with the union, intersection and complement operation.

$$\overline{A \cap B} = \overline{A} \cup \overline{B}$$

$$\overline{A \cup B} = \overline{A} \cap \overline{B}.$$

Table 1.1. Features of Crisp Set

(1) Involution	$\overline{\overline{A}} = A$
(2) Commutativity	$A \cup B = B \cup A$ $A \cap B = B \cap A$
(3) Associativity	$(A \cup B) \cup C = A \cup (B \cup C)$ $(A \cap B) \cap C = A \cap (B \cap C)$
(4) distributivity	$A \cap (B \cup C) = (A \cap B) \cup (A \cap C)$ $A \cup (B \cap C) = (A \cup B) \cap (A \cup C)$

Table 1.1. (cont')

(5) Idempotency	$A \cup A = A$ $A \cap A = A$
(6) Absorption	$A \cup (A \cap B) = A$ $A \cap (A \cup B) = A$
(7) Absorption by X and $\varnothing$	$A \cup X = X$ $A \cap \varnothing = \varnothing$
(8) Identity	$A \cup \varnothing = A$ $A \cap X = A$
(9) De Morgan's law	$\overline{A \cap B} = \overline{A} \cup \overline{B}$ $\overline{A \cup B} = \overline{A} \cap \overline{B}$
(10) Absorption of complement	$A \cup (\overline{A} \cap B) = A \cup B$ $A \cap (\overline{A} \cup B) = A \cap B$
(11) Law of contradiction	$A \cap \overline{A} = \varnothing$
(12) Law of excluded middle	$A \cup \overline{A} = X$

1.3.2 Convex Set

Definition (Convex set) The term convex is applicable to a set A in R^n (n-dimensional Euclidian vector space) if the followings are satisfied.

   i) Two arbitrary points s and r are defined in A.

$$r = (r_i \mid i \in N_n), \quad s = (s_i \mid i \in N_n).$$

      (N is a set of positive integers)

   ii) For arbitrary real number λ between 0 and 1, point t is involved in A
      where t is

$$t = (\lambda r_i + (1 - \lambda) s_i \mid i \in N_n). \quad \square$$

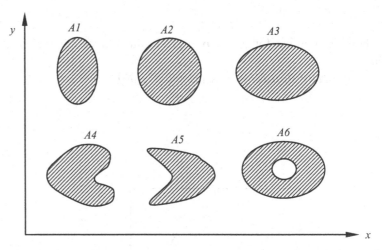

Fig. 1.1. Convex sets *A1, A2, A3* and non-convex sets *A4, A5, A6* in |R^2

In other wads, if every point on the line connecting two points *s* and *r* in *A* is also in *A*. (Fig 1.1) shows some examples of convex and non-convex sets

1.4 Definition of Fuzzy Set

1.4.1 Expression for Fuzzy Set

Membership function μ_A in crisp set maps whole members in universal set *X* to set {0,1}.

$$\mu_A : X \rightarrow \{0,\ 1\}.$$

Definition (Membership function of fuzzy set) In fuzzy sets, each elements is mapped to [0,1] by membership function.

$$\mu_A : X \rightarrow [0,\ 1]$$

where [0,1] means real numbers between 0 and 1 (including 0,1). □

Consequently, fuzzy set is 'vague boundary set' comparing with crisp set.

Example 1.2 (see Fig 1.2 and 1.3) show the difference between the crisp and fuzzy sets represented by membership functions, respectively. □

Example 1.3 Consider fuzzy set 'two or so'. In this instance, universal set X are the positive real numbers.

$$X = \{1, 2, 3, 4, 5, 6, \ldots\}$$

Membership function for A ='two or so' in this universal set X is given as follows:

$$\mu_A(1) = 0,\ \mu_A(2) = 1,\ \mu_A(3) = 0.5,\ \mu_A(4) = 0\ldots \quad \square$$

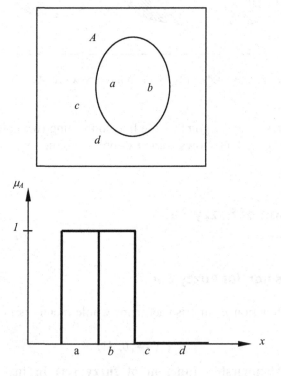

Fig. 1.2. Graphical representation of crisp set

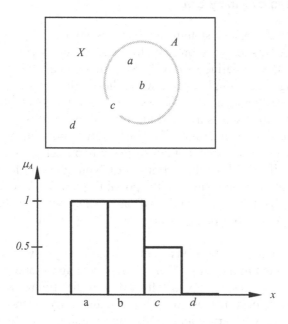

Fig. 1.3. Graphical representation of fuzzy set

Usually, if elements are discrete as the above, it is possible to have membership degree or grade as
$$A = \{(2, 1.0), (3, 0.5)\}$$
or
$$A = 1.0/2 + 0.5/3$$
be sure to notice that the symbol '+' implies not addition but union. More generally, we use
$$A = \{(x, \mu_A(x))\}$$
or
$$A = \sum_{i=1}^{n} \mu_A(x_i) / x_i .$$

Suppose elements are continuous, then the set can be represented as follows:
$$A = \int \mu_A(x) / x .$$

For the discrimination of fuzzy set with crisp set, the symbol $\tilde{A}$ is frequently used. However in this book, just notation A is used for it.

1.4.2 Examples of Fuzzy Set

Example 1.4 We consider statement "Jenny is young". At this time, the term "young" is vague. To represent the meaning of "vague" exactly, it would be necessary to define its membership function as in Fig 1.4. When we refer "young", there might be age which lies in the range [0,80] and we can account these "young age" in these scope as a continuous set.

The horizontal axis shows age and the vertical one means the numerical value of membership function. The line shows possibility (value of membership function) of being contained in the fuzzy set "young".

For example, if we follow the definition of "young" as in the figure, ten year-old boy may well be young. So the possibility for the "age ten" to join the fuzzy set of "young is 1. Also that of "age twenty seven" is 0.9. But we might not say young to a person who is over sixty and the possibility of this case is 0.

Now we can manipulate our last sentence to "Jenny is very young". In order to be included in the set of "very young", the age should be lowered and let us think the line is moved leftward as in the figure. If we define fuzzy set as such, only the person who is under forty years old can be included in the set of "very young". Now the possibility of twenty-seven year old man to be included in this set is 0.5.

That is, if we denote A= "young" and B="very young",
$$\mu_A(27) = 0.9, \quad \mu_B(27) = 0.5.$$

Example 1.5 Let's define a fuzzy set A ={real number near 0}. The boundary for set "real number near 0" is pretty ambiguous. The possibility of real number x to be a member of prescribed set can be defined by the following membership function.   □

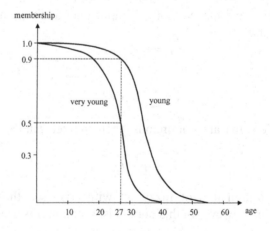

Fig. 1.4. Fuzzy sets representing "young" and "very young"

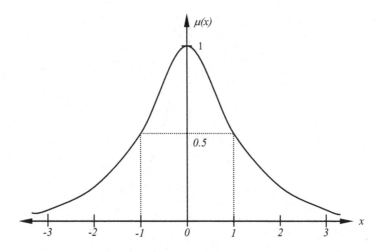

Fig. 1.5. Membership function of fuzzy set "real number near 0"

$$\mu_A(x) = \frac{1}{1+x^2}.$$

Fig 1.5 shows this membership function. We can also write the fuzzy set with the function.

$$A = \int \mu_A(x)/x \qquad \text{where } \mu_A(x) = \frac{1}{1+x^2}.$$

The membership degree of 1 is

$$\frac{1}{1+1^2} = 0.5$$

the possibility of 2 is 0.2 and that of 3 is 0.1. □

Example 1.6 Another fuzzy set A ={real number very near 0} can be defined and its membership function is

$$\mu_B(x) = \left(\frac{1}{1+x^2}\right)^2$$

the possibility of 1 is 0.25, that of 2 is 0.04 and of 3 is 0.01 (Fig 1.6).

By modifying the above function, it is able to denote membership function of fuzzy set A = {real number near a} as,

$$\mu_A(x) = \frac{1}{1+(x-a)^2}. □$$

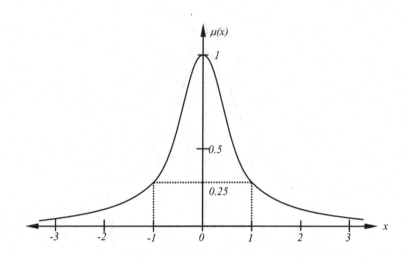

Fig. 1.6. Membership function for "real number very near to 0"

1.4.3 Expansion of Fuzzy Set

Definition (Type-n Fuzzy Set) The value of membership degree might include uncertainty. If the value of membership function is given by a fuzzy set, it is a *type-2* fuzzy set. This concept can be extended up to Type-n fuzzy set. □

Example 1.7 Consider set A= "adult". The membership function of this set maps whole age to "youth", "manhood" and "senior"(Fig 1.7). For instance, for any person x, y, and z,

$$\mu_A(x) = \text{"youth"}$$
$$\mu_A(y) = \text{"manhood"}$$
$$\mu_A(z) = \varnothing.$$

The values of membership for "youth" and "manhood" are also fuzzy sets , and thus the set "adult" is a type-2 fuzzy set.

The sets "youth" and "manhood" are type-1 fuzzy sets. In the same manner, if the values of membership function of "youth" and "manhood" are type-2, the set "adult" is type-3. □

Definition (Level-k fuzzy set) The term *"level-2 set"* indicates fuzzy sets whose elements are fuzzy sets (Fig 1.8). The term *"level-1 set"* is applicable to fuzzy sets whose elements are no fuzzy sets ordinary elements. In the same way, we can derive up to level-k fuzzy set.

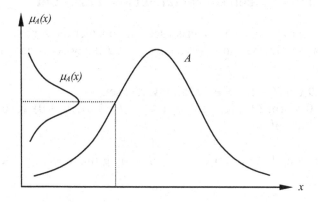

Fig. 1.7. Fuzzy Set of Type-2

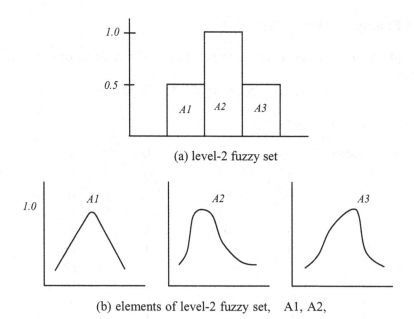

(a) level-2 fuzzy set

(b) elements of level-2 fuzzy set, A1, A2,

Fig. 1.8. Level-2 Fuzzy Set

Example 1.8 In the figure, there are 3 fuzzy set elements.

$$\mu_A(A_1) = 0.5$$
$$\mu_A(A_2) = 1.0$$
$$\mu_A(A_3) = 0.5. \quad \square$$

1.4.4 Relation between Universal Set and Fuzzy Set

If there are a universal set and a crisp set, we consider the set as a subset of the universal set. In the same way, we regard a fuzzy set A as a subset of universal set X.

Example 1.9 Let $X = \{a, b, c\}$ be a universal set.
$A_1 = \{(a, 0.5), (b, 1.0), (c, 0.5)\}$ and $A_2 = \{(a, 1.0), (b, 1.0), (c, 0.5)\}$ would be subsets of X.

$$A_1 \subseteq X, \quad A_2 \subseteq X.$$

The collection of these subsets of X (including fuzzy set) is called **power set** $P(X)$. □

1.5 Expanding Concepts of Fuzzy Set

1.5.1 Example of Fuzzy Set

Example 1.10 Consider a universal set X which is defined on the age domain.

$$X = \{5, 15, 25, 35, 45, 55, 65, 75, 85\}$$

Table 1.2. Example of fuzzy set

age(element)	infant	young	adult	senior
5	0	0	0	0
15	0	0.2	0.1	0
25	0	1	0.9	0
35	0	0.8	1	0
45	0	0.4	1	0.1
55	0	0.1	1	0.2
65	0	0	1	0.6
75	0	0	1	1
85	0	0	1	1

We can define fuzzy sets such as "infant", "young", "adult" and "senior" in X. The possibilities of each element of x to be in those four fuzzy sets are in Table 1.2

We can think of a set that is made up of elements contained in A. This set is called "**support**" of A.

$$Support(A) = \{x \in X \mid \mu_A(x) > 0\}.$$

The support of fuzzy set "young" is,

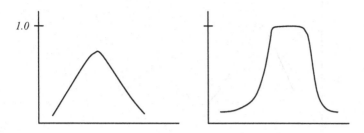

Fig. 1.9. Non-Normalized Fuzzy Set and Normalized Fuzzy Set

$$Support(youth) = \{15, 25, 35, 45, 55\}$$

and it is a crisp set. Certainly, the support of "infant" is empty set.

The maximum value of the membership is called "height". Suppose this "height" of some fuzzy sets is 1, this fuzzy set is "normalized". The sets "young", "adult" and "senior" are normalized (Fig 1.9).

Let's consider crisp set "teenager". This crisp set is clearly defined having elements only 10-19 in the universal set X. As you shall notice this set is a ***restricted set*** comparing with X. Similarly fuzzy set "young" is also a restricted set. When we apply a "***fuzzy restriction***" to universal set X in certain manner, we get a fuzzy set

1.5.2 α-Cut Set

Definition (α-cut set) The α-cut set A_α is made up of members whose membership is not less than α.
$$A_\alpha = \{x \in X \mid \mu_A(x) \geq \alpha\}$$
note that α is arbitrary. This α-cut set is a crisp set (Fig 1.10). □

Example 1.11 The α-cut set is derived from fuzzy set "young" by giving 0.2 to α
$$Young_{0.2} = \{12, 25, 35, 45\}$$
this means "the age that we can say youth with possibility not less than 0.2".

If α=0.4, $Young_{0.4} = \{25, 35, 45\}$
If α=0.8, $Young_{0.8} = \{25, 35\}$. □

When two cut sets A_α and $A_{\alpha'}$ exist and if $\alpha \leq \alpha'$ for them, then
$$A_\alpha \supseteq A_{\alpha'}$$
the relation $Young_{0.2} \supseteq Young_{0.8}$, for example, holds.

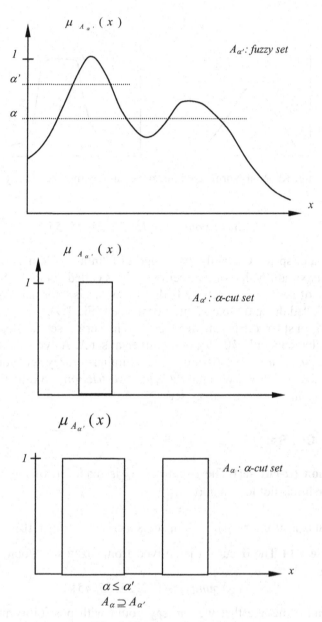

Fig. 1.10. α-cut set

Definition (Level set) The value α which explicitly shows the value of the membership function, is in the range of [0,1]. The "level set" is obtained by the α's. That is,

$$\Lambda_A = \{\alpha \mid \mu_A(x) = \alpha, \ \alpha \geq 0, \ x \in X\}. \quad \square$$

The level set of the above fuzzy set "young" is,

$$\Lambda_A = \{0, 0.1, 0.2, 0.4, 0.8, 1.0\}.$$

1.5.3 Convex Fuzzy Set

Definition (Convex fuzzy set) Assuming universal set X is defined in n-dimensional Euclidean Vector space $\Re^n$. If all the α- cut sets are convex, the fuzzy set with these α- cut sets is convex(Fig 1.11). In other words, if a relation

$$\mu_A(t) \geq \text{Min}[\mu_A(r), \mu_A(s)]$$

where $t = \lambda r + (1-\lambda)s$ $r, s \in \Re^n, \lambda \in [0,1]$

holds, the fuzzy set A is convex. □

Fig 1.12 shows a convex fuzzy set and Fig 1.13 describes a non-convex set.

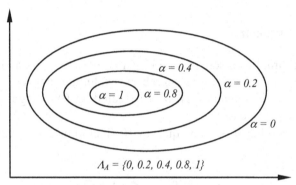

$$\Lambda_A = \{0, 0.2, 0.4, 0.8, 1\}$$

Fig. 1.11. Convex Fuzzy Set

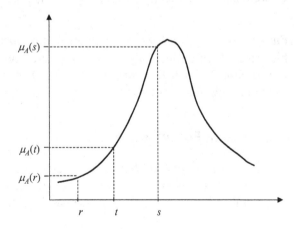

Fig. 1.12. Convex Fuzzy Set $\mu_A(t) \geq \mu_A(r)$

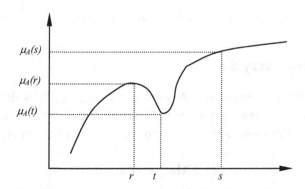

Fig. 1.13. Non-Convex Fuzzy Set $\mu_A(t) \not\geq \mu_A(r)$

1.5.4 Fuzzy Number

"Real number" implies a set containing whole real numbers and "Positive number" implies a set holding numbers excluding negative numbers. "Positive number less than equal to 10 (including 0)" suggests us a set having numbers from 0 to 10. So
A = "positive number less than equal to 10 (including 0)" = $\{x \mid 0 \leq x \leq 10, x \in \mathfrak{R}\}$
or

$$\mu_A(x) = 1 \text{ if } 0 \leq x \leq 10, x \in \mathfrak{R}$$
$$= 0 \text{ if } x < 0 \text{ or } x > 10$$

since the crisp boundary is involved, the outcome of membership function is 1 or 0.

Definition (Fuzzy number) If a fuzzy set is *convex* and *normalized*, and its membership function is defined in $\mathfrak{R}$ and *piecewise* continuous, it is called as *"fuzzy number"*. So fuzzy number (fuzzy set) represents a real number interval whose boundary is fuzzy(Fig 1.14).   □

1.5.5 The Magnitude of Fuzzy Set

In order to show the magnitude of fuzzy set, there are three ways of measuring the cardinality of fuzzy set. First, we can derive magnitude by summing up the membership degrees. It is *"scalar cardinality"*.

$$|A| = \sum_{x \in X} \mu_A(x).$$

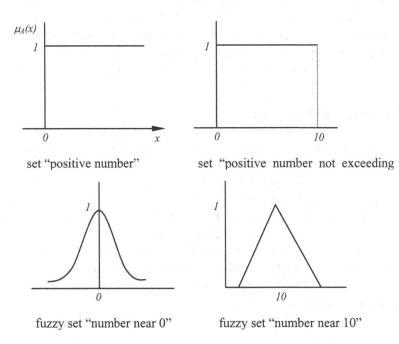

set "positive number"   set "positive number not exceeding

fuzzy set "number near 0"   fuzzy set "number near 10"

Fig. 1.14. Sets denoting intervals and fuzzy numbers

Following this method, the magnitude of fuzzy set "senior" (in the previous example) is,

$$|senior| = 0.1 + 0.2 + 0.6 + 1 + 1 = 2.9$$

Second, comparing the magnitude of fuzzy set A with that of universal set X can be an idea.

$$||A|| = \frac{|A|}{|X|}$$

This is called "relative cardinality". In the case of "senior",

$$|senior| = 2.9, |X| = 9$$
$$||senior|| = 2.9/9 = 0.32$$

Third method expresses the cardinality as fuzzy set.

Definition (Fuzzy cardinality) Let's try to get α-cut set (crisp set) A_α, of A. The number of elements is $|A_\alpha|$. In other words, the possibility for number of elements in A to be $|A_\alpha|$ is α. Then the membership degree of fuzzy cardinality $|A|$ is defined as,

$$\mu_{|A|}(|A_\alpha|) = \alpha, \quad \alpha \in \Lambda_A$$

where A_α is a α-cut set and Λ_A is a level set. □

Example 1.12 If we cut fuzzy set "senior" at $\alpha=0.1$, there are 5 elements in the α-cut set. $senior_{0.1} = \{45, 55, 65, 75, 85\}$, $/senior_{0.1}/ = 5$. In the same manner, there are 4 elements at $\alpha=0.2$, there are 3 at $\alpha=0.6$, there are 2 at $\alpha=1$. Therefore the fuzzy cardinality of "senior" is

$$|senior| = \{(5, 0.1), (4, 0.2), (3, 0.6), (2,1)\}. \quad \square$$

1.5.6 Subset of fuzzy set

Suppose there are two fuzzy sets A and B. When their degrees of membership are same, we say "A and B are equivalent". That is,

$$A = B \quad \text{iff} \quad \mu_A(x) = \mu_B(x), \quad \forall x \in X$$

If $\mu_A(x) \neq \mu_B(x)$ for any element, then $A \neq B$. If the following relation is satisfied in the fuzzy set A and B, A is a **subset** of B(Fig 1.15).

$$\mu_A(x) \leq \mu_B(x), \forall x \in X$$

This relation is expressed as $A \subseteq B$. We call that A is a subset of B. In addition, if the next relation holds, A is a **proper subset** of B.

$$\mu_A(x) < \mu_B(x), \forall x \in X$$

This relation can be written as

$$A \subset B \quad \text{iff} \quad A \subseteq B \text{ and } A \neq B.$$

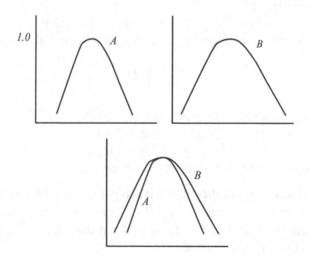

Fig. 1.15. Subset $A \subset B$

1.6 Standard Operation of Fuzzy Set

1.6.1 Complement

We can find complement set of fuzzy set A likewise in crisp set. We denote the complement set of A as $\overline{A}$. Membership degree can be calculated as following.

$$\mu_{\overline{A}}(x) = 1 - \mu_A(x), \quad \forall x \in X.$$

If we calculate the complement set of "adult" as $\overline{A}$, we may have

$$\overline{A} = \{(5, 1), (15, 0.9), (25, 0.1)\}.$$

1.6.2 Union

Membership value of member x in the union takes the greater value of membership between A and B

$$\mu_{A \cup B}(x) = \text{Max}[\mu_A(x), \mu_B(x)], \quad \forall x \in X.$$

Of course, A and B are subsets of $A \cup B$, and union of "young" and "adult" is (Table 1.2),

"young"$\cup$"adult" = {(15,0.2), (25,1), (35,1), (45,1), (55,1), (65,1), (75,1), (85,1)}.

1.6.3 Intersection

Intersection of fuzzy sets A and B takes smaller value of membership function between A and B.

$$\mu_{A \cap B}(x) = \text{Min}[\mu_A(x), \mu_B(x)], \quad \forall x \in X.$$

Intersection $A \cap B$ is a subset of A or B. For instance, the intersection of "young" and "adult" is (Table 1.2),

"young" $\cap$ "adult" = {(15, 0.1), (25, 0.9), (35, 0.8), (45, 0.4), (55, 0.1)}.

We have seen that complement set, union and intersection sets are applicable even if membership function is restricted to 0 or 1 (*i.e.* crisp set). We will see more about fuzzy operations in the next chapter.

[SUMMARY]

☐ Operation of crisp set
  – Complement set
  – Union set
  – Intersection set

☐ Convex set
  $t \in A$ where
  $t = \lambda r + (1 - \lambda)s, \quad 0 \le \lambda \le 1$
  $r, \quad s \in A$

☐ Expression of membership degree of fuzzy set A
  – Membership function $\mu_A(x), \quad 0 \le \mu_A(x) \le 1$

☐ Expression of fuzzy set
  $$A = \{(x, \mu_A(x))\}$$
  $$A = \mu_A(x_1) / x_1 + \mu_A(x_2) / x_2 + \cdots$$
  $$A = \sum_{i=1,n} \mu_A(x_i) / x_i$$
  $$A = \int \mu_A(x) / x$$

☐ Extension of fuzzy set
  – Type-n fuzzy set
  – Level-k fuzzy set

☐ α-cut set
  – Crisp set whose elements have at least α degree of membership in fuzzy set A
  $$A_\alpha = \{x \in X \mid \mu_A(x) \ge \alpha\}$$

☐ Level set
  – Set gathered degrees of fuzzy set
  $$\{\alpha \mid \mu_A(x) = \alpha, \alpha \ge 0, x \in X\}$$

☐ Convex fuzzy set
  – For $r, s \in \Re^n$
  $$\mu_A(t) \ge Min[\mu_A(r), \mu_A(s)]$$

where $t = \lambda r + (1-\lambda)s \quad \lambda \in [0, 1]$

☐ Fuzzy number
- Convex fuzzy set.
- Normalized set.
- Continuous membership function

☐ Cardinality of fuzzy set
- Scalar cardinality
$$| A | = \sum \mu_A(x)$$
- Relative cardinality
$$\| A \| = | A | / | X |$$
- Fuzzy cardinality : denote the size of fuzzy set as $/A/$.
$$\mu_{|A|}(| A_\alpha |) = \alpha$$
$\alpha \in \Lambda_A$ (level set), A_α is α-cut set

☐ Fuzzy subset $A \supseteq B$
$$\mu_A(x) \geq \mu_B(x), \quad \forall x \in X$$
- Fuzzy proper subset $A \supset B$
$$\mu_A(x) > \mu_B(x), \quad \forall x \in X$$

[EXERCISES]

1.1 Show that the following sets satisfy the law of contradiction and law of excluded middle.

$X = \{a, b, c, d, e, f, g\}$

$A = \{a, b, c, d\}$

1.2 $A = \sum_{i=1,n} \mu_A(x_i)/x_i$ is an another form of representation of fuzzy set.

Represent the following fuzzy sets by this form.

a) $A = \{(2, 1.0), (3, 0.4), (4, 0.5)\}$

b) $B = \{(a, \mu_B(a)), (b, \mu_B(b)), (c, \mu_B(c)), (d, \mu_B(d))\}$

1.3 Consider the fuzzy sets : short, middle, tall

cm	short	middle	tall
140	1	0	0
150	1	0	0
160	0.9	0.1	0
170	0.7	1	0
180	0.3	0.8	0.3
190	0	0	1

a) Compare the support of each set.

b) What is the normalized fuzzy set?

c) Find the level set of each set.

d) Compare α-cut set of each set where α=0.5 and α=0.3.

1.4 Determine whether the following fuzzy sets are convex or not.

a) $A = \int \mu_A(x)/x$ where $\mu_A(x) = 1/(1+x^2)$

b) $B = \int \mu_B(x)/x$ where $\mu_B(x) = 1/(1+10x)^{1/2}$

1.5 Prove that all the α–cuts of any fuzzy set A defined on R^n are convex if and only if

$$\mu_A(\lambda r + (1-\lambda)s) \geq Min[\mu_A(r), \mu_A(s)]$$

such that $r, s \in \Re^n$, $\lambda \in [0,1]$

1.6 Compute the scalar cardinality and the fuzzy cardinality for each of the following fuzzy set.

a) A = {(x, 0.4), (y, 0.5), (z, 0.9), (w, 1)}
b) B = {0.5/u + 0.8/v + 0.9/w + 0.1/x}
c) $C = \sum \mu_C(x)/x$ where $\mu_C(x) = (x/(x+1))^2$ $x \in \{0,1,2,\cdots 10\}$

1.7 Show the following set is convex.
$$\mu_A(x) = \begin{cases} 0 & x \le 10 \\ (1+(x-10)^{-2})^{-1} & x > 10 \end{cases}$$

1.8 Determine α-cut sets of the above set for α=0.5, 0.8 and 0.9.

Chapter 2. THE OPERATION OF FUZZY SET

In the previous section, we have studied complement, union and intersection operations of fuzzy sets. In this chapter various operations of fuzzy sets are introduced more formally. The concepts of disjunctive sum, distance, difference, conorm and t – conorm operators are also given.

2.1 Standard Operations of Fuzzy Set

Complement set $\overline{A}$, union $A \cup B$, and intersection $A \cap B$ represent the standard operations of fuzzy theory and are arranged as,

$$\mu_{\overline{A}}(x) = 1 - \mu_A(x)$$

$$\mu_{A \cup B}(x) = \text{Max}[\mu_A(x), \mu_B(x)]$$

$$\mu_{A \cap B}(x) = \text{Min}[\mu_A(x), \mu_B(x)]$$

We can find out that the above three operations are generalizations of those in crisp set. In the next sections, we'll see variety of proposed operators besides these operations. These various operators must satisfy necessary conditions, and they are put into use in diverse application fields.

Among numerous operators, since complement, Max and Min operators are fundamental and simple, many fuzzy theories are develop based upon these operators. Table 2.1 includes characteristics of the standard operators. Note that the following two characteristics of crisp set operators does not hold here.

$$\text{law of contradiction} \qquad A \cap \overline{A} = \varnothing$$

$$\text{law of excluded middle} \qquad A \cup \overline{A} = X$$

The reason for this occurrence is that the boundary of complement of A is ambiguous.

2.2 Fuzzy Complement

2.2.1 Requirements for Complement Function

Complement set $\overline{A}$ of set A carries the sense of negation. Complement set may be defined by the following function C.

Table 2.1. Characteristics of standard fuzzy set operators

(1) Involution	$\overline{\overline{A}} = A$
(2) Commutativity	$A \cup B = B \cup A$ $A \cap B = B \cap A$
(3) Associativity	$(A \cup B) \cup C = A \cup (B \cup C)$
(4) Distributivity	$(A \cap B) \cap C = A \cap (B \cap C)$ $A \cap (B \cup C) = (A \cap B) \cup (A \cap C)$ $A \cup (B \cap C) = (A \cup B) \cap (A \cup C)$
(5) Idempotency	$A \cup A = A$ $A \cap A = A$
(6) Absorption	$A \cup (A \cap B) = A$ $A \cap (A \cup B) = A$
(7) Absorption by X and $\varnothing$	$A \cup X = X$ $A \cap \varnothing = \varnothing$
(8) Identity	$A \cup \varnothing = A$ $A \cap X = A$
(9) De Morgan's law	$\overline{A \cap B} = \overline{A} \cup \overline{B}$ $\overline{A \cup B} = \overline{A} \cap \overline{B}$

(10) Equivalence formula

$$(\overline{A} \cup B) \cap (A \cup \overline{B}) = (\overline{A} \cap \overline{B}) \cup (A \cap B)$$

(11) Symmetrical difference formula

$$(\overline{A} \cap B) \cup (A \cap \overline{B}) = (\overline{A} \cup \overline{B}) \cap (A \cup B)$$

$$C : [0,1] \to [0,1]$$

Complement function C is designed to map membership function $\mu_A(x)$ of fuzzy set A to $[0,1]$ and the mapped value is written as $C(\mu_A(x))$. To be a fuzzy complement function, two axioms should be satisfied.

(Axiom C1) $C(0) = 1$, $C(1) = 0$ (boundary condition)

(Axiom C2) $a,b \in [0,1]$

   if $a < b$, then $C(a) \geq C(b)$ (monotonic nonincreasing)

Symbols a and b stand for membership value of member x in A. For example, when $\mu_A(x) = a$, $\mu_A(y) = b$, $x, y \in X$, if $\mu_A(x) < \mu_A(y)$, $C(\mu_A(x)) \geq C(\mu_A(y))$.

C1 and C2 are fundamental requisites to be a complement function. These two axioms are called "axiomatic skeleton". For particular purposes, we can insert additional requirements.

(Axiom C3) C is a continuous function.

(Axiom C4) C is involutive.

   $C(C(a)) = a$ for all $a \in [0,1]$

2.2.2 Example of Complement Function

Above four axioms hold in standard complement operator

$$C(\mu_A(x)) = 1 - \mu_A(x) \text{ or } \mu_{\bar{A}}(x) = 1 - \mu_A(x)$$

this standard function is shown in Fig 2.1, and it's visual representation is given in (Fig 2.2.)

The following is a complement function satisfying only the axiomatic skeleton (Fig 2.3).

$$C(a) = \begin{cases} 1 & \text{for } a \leq t \\ 0 & \text{for } a > t \end{cases}$$

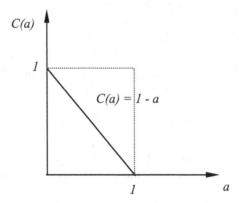

Fig. 2.1. Standard complement set function

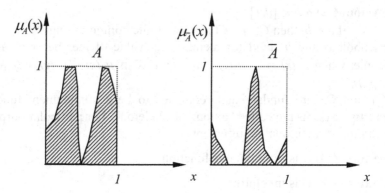

Fig. 2.2. Illustration of standard complement set function

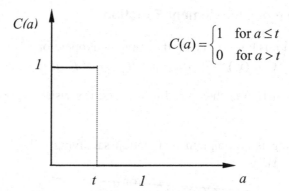

$$C(a) = \begin{cases} 1 & \text{for } a \le t \\ 0 & \text{for } a > t \end{cases}$$

Fig. 2.3. Example of complement set function

Note that it does not hold C3 and C4. Again, the following complement function is continuous (C3) but not holds C4 (Fig 2.4).

$$C(a) = 0.5(1 + \cos \pi a)$$

When $a = 0.33$, $C(0.33) = 0.75$ in this function. However since $C(0.75) = 0.15 \ne 0.33$, C4 does not hold now. One of the popular complement functions, is Yager's function as in the following :

$$C_w(a) = (1 - a^w)^{1/w} \qquad \text{where } w \in (-1, \infty)$$

The shape of the function is dependent on parameter (Fig 2.5). When $w = 1$, the Yager's function becomes the standard complement function $C(a) = 1 - a$.

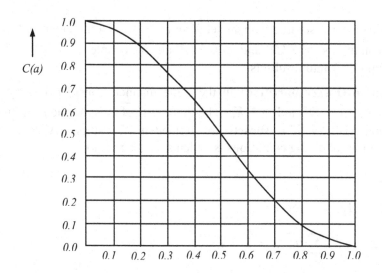

Fig.2.4. Continuous fuzzy complement function $C(a) = 1/2(1+\cos \pi a)$

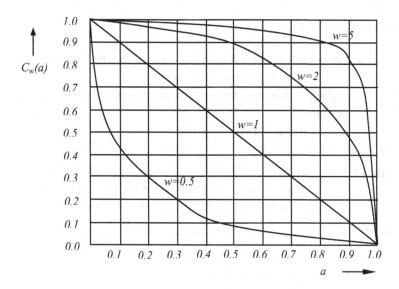

Fig. 2.5. Yager complement function

2.2.3 Fuzzy Partition

Let A be a crisp set in universal set X and $\overline{A}$ be a complement set of A. The conditions $A \neq \varnothing$ and $A \neq X$ result in couple $(A, \overline{A})$ which decomposes X into 2 subsets.

Definition (Fuzzy partition) In the same manner, consider a fuzzy set satisfying $A \neq \varnothing$ and $A \neq X$.. The pair $(A, \overline{A})$ is defined as fuzzy partition. Usually, if m subsets are defined in X, m-tuple $(A_1, A_2, ..., A_n)$ holding the following conditions is called a fuzzy partition.

i) $\forall i, \quad A_i \neq \varnothing$

ii) $A_i \cap A_j = \varnothing$ for $i \neq j$

iii) $\forall x \in X, \quad \sum_{i=1}^{m} \mu_{A_i}(x) = 1$ □

2.3 Fuzzy Union

2.3.1 Axioms for Union Function

In general sense, union of A and B is specified by a function of the form.
$$U:[0,1] \times [0,1] \to [0,1]$$
this union function calculates the membership degree of union $A \cup B$ from those of A and B.
$$\mu_{A \cup B}(x) = U[\mu_A(x), \mu_B(x)]$$
this union function should obey next axioms.

(Axiom U1) $U(0,0) = 0, U(0,1) = 1, U(1,0) = 1, U(1,1) = 1$
so this union function follows properties of union operation of crisp sets (boundary condition).

(Axiom U2) $U(a,b) = U(b,a)$ Commutativity holds.

(Axiom U3)   If $a \leq a'$ and $b \leq b'$, $U(a, b) \leq U(a', b')$ Function U is a monotonic function.

(Axiom U4) $U(U(a, b), c) = U(a, U(b, c))$ Associativity holds.
the above four statements are called as "axiomatic skeleton". It is often to restrict the class of fuzzy unions by adding the following axioms.

(Axiom U5)   Function U is continuous.

(Axiom U6) $U(a, a) = a$ (idempotency)

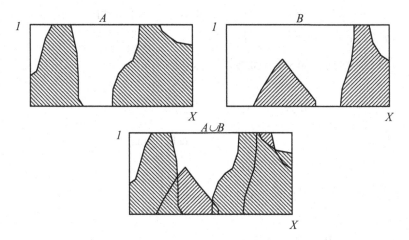

Fig. 2.6. Visualization of standard union operation

2.3.2 Examples of Union Function

The standard operator Max is treading on those six axioms.
$$U[\mu_A(x),\ \mu_B(x)] = Max[\mu_A(x),\ \mu_B(x)]$$
or
$$\mu_{A\cup B}(x) = Max[\mu_A(x),\ \mu_B(x)]$$
visualizing the standard union operation leads to Fig 2.7.

Yager's union function holds all axioms except U6.
$$U_w(a,b) = Min[1, (a^w + b^w)^{1/w}], \quad \text{where} \quad w \in (0, \infty)$$
the shape of Yager function varies with parameter w. For instance,
$w = 1$ leads to.
$$U_1(a, b) = Min[1, a+b]$$
$w = 2$ leads to
$$U_2(a, b) = Min[1, \sqrt{a^2 + b^2}\ ]$$

What if w increases? Supposing $w \to \infty$, Yager union function is transformed into the standard union function.
$$\lim_{w \to \infty} Min[1, (a^w + b^w)^{1/w}] = Max(a,b)$$

there are some examples of Yager function for $w = 1$, 2 and ∞ *in* (Fig 2.7). We know that the union operation of crisp sets is identical to the OR logic. It is easy to see the relation is also preserved here. For example, let's reconsider the example in the previous chapter. If set A be "young" and B

a \ b	0	0.25	0.5
1	1	1	1
0.75	0.75	1	1
0.25	0.25	0.5	0.75

$U_1(a,b) = Min[1, a+b]$

$w = 1$

a \ b	0	0.25	0.5
1	1	1	1
0.75	0.75	0.79	0.9
0.25	0.25	0.35	0.55

$U_2(a,b) = Min[1, \sqrt{a^2 + b^2}]$

$w = 2$

a \ b	0	0.25	0.5
1	1	1	1
0.75	0.75	0.75	0.75
0.25	0.25	0.25	0.5

$U_\infty(a,b) = Max[a, b]$

$w \to \infty$

Fig. 2.7. Yager's union function

"senior", the union of A and B is "young or senior". In the sense of meaning, the union and OR logic are completely identical.

2.3.3 Other Union Operations

(1) Probabilistic sum $A \hat{+} B$ (Algebraic sum)
Fuzzy union $A \hat{+} B$ is defined as,

$$\forall x \in X, \quad \mu_{A \hat{+} B}(x) = \mu_A(x) + \mu_B(x) - \mu_A(x)\mu_B(x)$$

It follows commutativity, associativity, identity, and De Morgan's law. This operator holds also the following :

$$A \hat{+} X = X$$

(2) Bounded sum $A \oplus B$ (Bold union)

$$\forall x \in X, \quad \mu_{A \oplus B}(x) = Min[1, \mu_A(x) + \mu_B(x)]$$

This operator is identical to Yager function at $w = 1$. Commutativity, associativity, identity, and De Morgan's Law are perfected, and it has relations,

$$A \oplus X = X$$
$$A \oplus \overline{A} = X$$

but it does not idempotency and distributivity at absorption.

(3) Drastic sum $A \mathbin{\dot\cup} B$

Drastic sum is defined as follows :

$$\forall x \in X, \quad \mu_{A \mathbin{\dot\cup} B}(x) = \begin{cases} \mu_A(x), & \text{when } \mu_B(x) = 0 \\ \mu_B(x), & \text{when } \mu_A(x) = 0 \\ 1, & \text{for others} \end{cases}$$

(4) Hamacher's sum $A \cup B$

$$\forall x \in X, \quad \mu_{A \cup B}(x) = \frac{\mu_A(x) + \mu_B(x) - (2 - \gamma)\mu_A(x)\mu_B(x)}{1 - (1 - \gamma)\mu_A(x)\mu_B(x)}, \quad \gamma \ge 0$$

2.4 Fuzzy Intersection

2.4.1 Axioms for Intersection Function

In general sense, intersection $A \cap B$ is defined by the function I.
$$I{:}[0,1] \times [0,1] \to [0,1]$$
The argument of this function shows possibility for element x to be involved in both fuzzy sets A and B.

$$\mu_{A \cap B}(x) = I[\mu_A(x), \mu_B(x)]$$

intersection function holds the following axioms .

(Axiom I1) $I(1, 1) = 1$, $I(1, 0) = 0$, $I(0, 1) = 0$, $I(0, 0) = 0$
Function I follows the intersection operation of crisp set (boundary condition).

(Axiom I2) $I(a, b) = I(b, a)$, Commutativity holds.

(Axiom I3)   If $a \le a'$ and $b \le b'$, $I(a, b) \le I(a', b')$, Function I is a monotonic function.

(Axiom I4) $I(I(a, b), c) = I(a, I(b, c))$, Associativity holds.
Just like in the union function, these four axioms are the axiomatic skeleton, and the following two axioms can be added.

(Axiom I5)   *I* is a continuous function

(Axiom I6)   *I(a, a) = a*, *I* is idempotency.

2.4.2 Examples of Intersection

Standard fuzzy intersection completes the above 6 axioms.

$$I[\mu_A(x), \mu_B(x)] = Min[\mu_A(x), \mu_B(x)]$$

or

$$\mu_{A \cap B}(x) = Min[\mu_A(x), \mu_B(x)]$$

Visualizing this standard intersection results Fig 2.8 using fuzzy sets *A* and *B* in (Fig 2.6.)

Considering Yager function as we did in union function, it steps on all axioms but for I6.

$$I_w(a,b) = 1 - Min[1, ((1-a)^w + (1-b)^w)^{1/w}],$$

where w ∈ (0, ∞)

The shape varies depending on *w*,
if w = 1

$$I_1(a, b) = 1 - Min[1, 2-a-b]$$

if w = 2

$$I_2(a, b) = 1 - Min[1, \sqrt{(1-a)^2 + (1-b)^2}]$$

and so forth.

What if *w* approaches infinity (*w*→∞)? The answer to this question is 'Yager function converges to the standard intersection function'.

$$\lim_{w \to \infty} (1 - Min[1, ((1-a)^w + (1-b)^w)^{1/w}]) = Min(a,b)$$

Note that intersection and *AND* logic are equivalent. For instance, consider two fuzzy sets "young" and "senior". Intersection for these is "person who is at once young and senior".

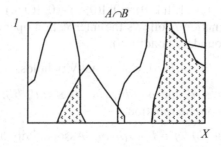

Fig. 2.8. Visualization of standard fuzzy intersection set

Example 2.1 Take Yager function at $w = 1$ for example. Providing $a = 0.4$ and $b = 0.6$, then

$$I_1 = 1 - \text{Min}[1, 2-(a+b)] = 1 - \text{Min}[1, 2-1] = 1 - 1 = 0$$

this time let $a = 0.5$ and $b = 0.6$, then $a+b=1.1$

$$I_1 = 1 - \text{Min}[1, 2-1.1] = 1 - \text{Min}[1, 0.9] = 1 - 0.9 = 0.1$$

take $a = 0.3$ and $b = 0.6$ for example. If $w \to \infty$, the intersection is reduced to,

$$I_\infty(a, b) = \text{Min}[0.3, 0.6] = 0.3$$

but when $w = 1$, $a + b = 0.9$; hence,

$$I_1(a, b) = 1 - \text{Min}[1, 2 - 0.9] = 1 - \text{Min}[1, 1.1] = 1 - 1 = 0 \quad \square$$

There are some more examples of Yager function in (Fig2.9)

b a	0	0.25	0.5
1	0	0.25	0.5
0.75	0	0	0.25
0.25	0	0	0

$I_1(a,b) = 1 - \text{Min}[1, 2-a-b]$

$w = 1$

b a	0	0.25	0.5
1	0	0.25	0.5
0.75	0	0.21	0.44
0.25	0	0	0.1

$I_2(a,b) = 1 - \text{Min}[1, \sqrt{(1-a)^2 + (1-b)^2}\,]$

$w = 2$

b a	0	0.25	0.5
1	0	0.25	0.5
0.75	0	0.25	0.5
0.25	0	0.25	0.25

$I_\infty(a,b) = \text{Min}[a, b]$

$w \to \infty$

Fig. 2.9. Yager's intersection function

2.4.3 Other Intersection Operations

(1) Algebraic product $A \bullet B$ (probabilistic product)
$$\forall x \in X, \quad \mu_{A \bullet B}(x) = \mu_A(x) \bullet \mu_B(x)$$
Operator $\bullet$ is obedient to rules of commutativity, associativity, identity, and De Morgan's law.

(2) Bounded product $A \odot B$ (Bold intersection)

This operator is defined as,

$$\forall x \in X, \quad \mu_{A \odot B}(x) = \text{Max}[0, \mu_A(x) + \mu_B(x) - 1]$$

and is identical to Yager intersection function with $w = 1$,
$$I_1(a, b) = 1 - \text{Min}[1, 2 - a - b]$$
commutativity, associativity, identity, and De Morgan's law hold in this operator . The following relations

$$A \odot \varnothing = \varnothing$$

$$A \odot \overline{A} = \varnothing$$

are also satisfied, but not idempotency, distributability, and absorption.

(3) Drastic product $A \cap B$

$$\forall x \in X, \quad \mu_{A \cap B}(x) = \begin{cases} \mu_A(x), & \text{when } \mu_A(x) = 1 \\ \mu_B(x), & \text{when } \mu_B(x) = 1 \\ 0, & \text{when } \mu_A(x), \ \mu_B(x) < 1 \end{cases}$$

(4) Hamacher's intersection $A \cap B$
$\forall x \in X,$

$$\mu_{A \cap B}(x) = \frac{\mu_A(x)\mu_B(x)}{\gamma + (1-\gamma)(\mu_A(x) + \mu_B(x) - \mu_A(x)\mu_B(x))}, \quad \gamma \geq 0$$

2.5 Other Operations in Fuzzy Set

2.5.1 Disjunctive Sum

Disjunctive sum is the name of operation corresponding "exclusive OR" logic. And it is expressed as the following (Fig 2.10)

$$A \oplus B = (A \cap \overline{B}) \cup (\overline{A} \cap B)$$

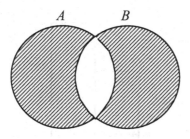

Fig. 2.10. Disjunctive sum of two sets

Definition (Simple disjunctive sum) By means of fuzzy union and fuzzy intersection, definition of the disjunctive sum in fuzzy set is allowed just like in crisp set.

$$\mu_{\bar{A}}(x) = 1 - \mu_A(x), \quad \mu_{\bar{B}}(x) = 1 - \mu_B(x)$$

$$\mu_{A \cap \bar{B}}(x) = Min[\mu_A(x), 1 - \mu_B(x)]$$

$$\mu_{\bar{A} \cap B}(x) = Min[1 - \mu_A(x), \mu_B(x)]$$

$$A \oplus B = (A \cap \bar{B}) \cup (\bar{A} \cap B), \quad \text{then}$$

$$\mu_{A \oplus B}(x) = Max\{Min[\mu_A(x), 1 - \mu_B(x)], \quad Min[1 - \mu_A(x), \mu_B(x)]\} \ \square$$

Example 2.2 Here goes procedures obtaining disjunctive sum of A and B (Fig 2.12).

$$A = \{(x_1, 0.2), (x_2, 0.7), (x_3, 1), (x_4, 0)\}$$
$$B = \{(x_1, 0.5), (x_2, 0.3), (x_3, 1), (x_4, 0.1)\}$$
$$\bar{A} = \{(x_1, 0.8), (x_2, 0.3), (x_3, 0), (x_4, 1)\}$$
$$\bar{B} = \{(x_1, 0.5), (x_2, 0.7), (x_3, 0), (x_4, 0.9)\}$$
$$A \cap \bar{B} = \{(x_1, 0.2), (x_2, 0.7), (x_3, 0), (x_4, 0)\}$$
$$\bar{A} \cap B = \{(x_1, 0.5), (x_2, 0.3), (x_3, 0), (x_4, 0.1)\}$$

and as a consequence,

$$A \oplus B = (A \cap \bar{B}) \cup (\bar{A} \cap B) = \{(x_1, 0.5), (x_2, 0.7), (x_3, 0), (x_4, 0.1)\} \quad \square$$

Definition (Disjoint sum) The key idea of "exclusive OR" is elimination of common area from the union of A and B. With this idea, we can define an operator $\triangle$ for the exclusive OR disjoint sum as follows.

$$\mu_{A \triangle B}(x) = |\mu_A(x) - \mu_B(x)| \quad \square$$

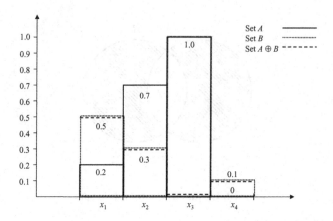

Fig. 2.11. Example of simple disjunctive sum

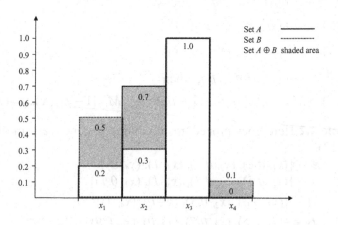

Fig. 2.12. Example of disjoint sum (exclusive OR sum)

Example 2.3 If we reconsider the previous example, we have (Fig 2.13)

$$A = \{(x_1, 0.2), (x_2, 0.7), (x_3, 1), (x_4, 0)\}$$
$$B = \{(x_1, 0.5), (x_2, 0.3), (x_3, 1), (x_4, 0.1)\}$$
$$A \triangle B = \{(x_1, 0.3), (x_2, 0.4), (x_3, 0), (x_4, 0.1)\} \quad \square$$

2.5.2 Difference in Fuzzy Set

The difference in crisp set is defined as follows (Fig 2.13.)

$$A - B = A \cap \overline{B}$$

In fuzzy set, there are two means of obtaining the difference
(1) Simple difference

Example 2.4 By using standard complement and intersection operations, the difference operation would be simple. If we reconsider the previews example, $A - B$ would be, (Fig 2.14)

$$A = \{(x_1, 0.2), (x_2, 0.7), (x_3, 1), (x_4, 0)\}$$
$$B = \{(x_1, 0.5), (x_2, 0.3), (x_3, 1), (x_4, 0.1)\}$$
$$\overline{B} = \{(x_1, 0.5), (x_2, 0.7), (x_3, 0), (x_4, 0.9)\}$$
$$A - B = A \cap \overline{B} = \{(x_1, 0.2), (x_2, 0.7), (x_3, 0), (x_4, 0)\} \quad \square$$

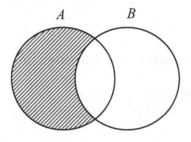

Fig. 2.13. Difference $A - B$

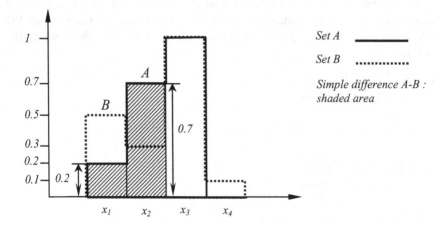

Fig. 2.14. Simple difference $A - B$

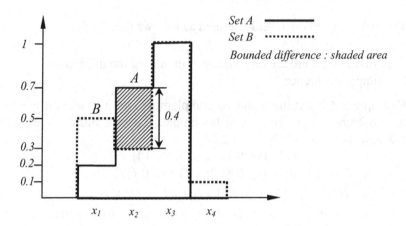

Fig. 2.15. Bounded difference $A \theta B$

(2) Bounded difference

Definition(Bounded difference)   For novice-operator θ, we define the membership function as,

$$\mu_{A\theta B}(x) = \text{Max}[0, \mu_A(x) - \mu_B(x)] \quad \square$$

By this definition , bounded difference of preceeding two fuzzy sets is as follows (Fig 2.15).

$$A \theta B = \{(x_1, 0), (x_2, 0.4), (x_3, 0), (x_4, 0)\}$$

2.5.3 Distance in Fuzzy Set

The concept 'distance' is designated to describe the difference. But it has different mathematical measure from the 'difference' introduced in the previews section (Fig 2.16). Measures for distance are defined in the following.

(1) Hamming distance
This concept is marked as,

$$d(A, B) = \sum_{i=1, x_i \in X}^{n} |\mu_A(x_i) - \mu_B(x_i)|$$

Example 2.5 Following A and B for instance,

$$A = \{(x_1, 0.4), (x_2, 0.8), (x_3, 1), (x_4, 0)\}$$
$$B = \{(x_1, 0.4), (x_2, 0.3), (x_3, 0), (x_4, 0)\}$$

Hamming distance; $d(A, B)$,

$$d(A, B) = |0| + |0.5| + |1| + |0| = 1.5 \quad \square$$

Hamming distance contains usual mathematical senses of 'distance'.
(1) $d(A, B) \geq 0$
(2) $d(A, B) = d(B, A)$ commutativity
(3) $d(A, C) \leq d(A, B) + d(B, C)$ transitivity
(4) $d(A, A) = 0$
Assuming n elements in universal set X; i.e., $|X| = n$, the relative Hamming distance is,

$$\delta(A, B) = \frac{1}{n} d(A, B)$$

might rename Hamming distance as 'symmetrical distance' and written as below by using operator ∇,

$$\forall x \in X, \quad \mu_{A\nabla B}(x) = |\mu_A(x) - \mu_B(x)|$$

This operator doesn't hold distributivity. In addition, 'disjoint sum' using operator Δ introduced in section 2.5.1 may be applied to this symmetrical distance.

(2) Euclidean distance
This novel term is arranged as,

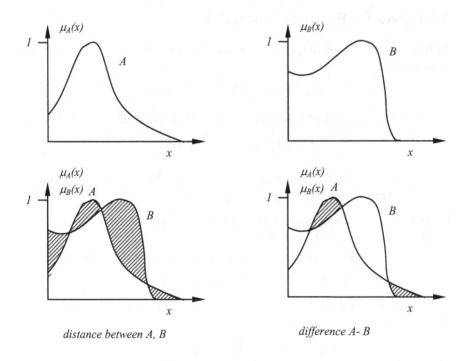

distance between A, B   difference A- B

Fig. 2.16. Distance and difference of fuzzy set

$$e(A, B) = \sqrt{\sum_{i=1}^{n} (\mu_A(x) - \mu_B(x))^2}$$

Example 2.6 Euclidean distance between sets A and B used for the previous Hamming distance is

$$e(A, B) = \sqrt{0^2 + 0.5^2 + 1^2 + 0^2} = \sqrt{1.25} = 1.12$$

and relative Euclidean distance is

$$\varepsilon(A, B) = \frac{e(A,B)}{\sqrt{n}}$$

(3) Minkowski distance

$$d_w(A, B) = \left(\sum_{x \in X} |\mu_A(x) - \mu_B(x)|^w \right)^{1/w}, \qquad\qquad w \in [1, \infty]$$

Generalizing Hamming distance and Euclidean distance results in Minkowski distance. It becomes the Hamming distance for $w = 1$ while the Euclidean distance for $w = 2$.

2.5.4 Cartesian Product of Fuzzy Set

Definition (Power of fuzzy set) Second power of fuzzy set A is defined as follows :

$$\mu_{A^2}(x) = [\mu_A(x)]^2, \quad \forall x \in X$$

Similarly m^{th} power of fuzzy set A^m may be computed as,

$$\mu_{A^m}(x) = [\mu_A(x)]^m, \quad \forall x \in X \quad \square$$

This operator is frequently applied when dealing with the linguistic hedge in expression of fuzzy set in chapter 8.

Definition (Cartesian product) Cartesian product applied to multiple fuzzy sets can be defined as follws.

Denoting $\mu_{A_1}(x)$, $\mu_{A_2}(x)$, ..., $\mu_{A_n}(x)$ as membership functions of $A_1, A_2, ..., A_n$ for $\forall x_1 \in A_1$, $x_2 \in A_2$, ..., $x_n \in A_n$.

then, the probability for n-tuple $(x_1, x_2, ..., x_n)$ to be involved in fuzzy set $A_1 \times A_2 \times ... \times A_n$ is,

$$\mu_{A_1 \times A_2 \times ... \times A_n}(x_1, x_2, ..., x_n) = \text{Min}[\mu_{A_1}(x_1), ..., \mu_{A_n}(x_n)] \quad \square$$

2.6 t-norms and t-conorms

There are two types of operators in fuzzy sets: t-norms and t-conorms. They are often called as triangular-norm and triangular-conorm respectively.

2.6.1 Definitions for t-norms and t-conorms

Definition (t-norm)

$T : [0,1] \times [0,1] \rightarrow [0,1]$

$\forall x, y, x', y', z \in [0,1]$

i) $T(x, 0) = 0, T(x, 1) = x$: boundary condition

ii) $T(x, y) = T(y, x)$: commutativity

iii) $(x \leq x', y \leq y') \rightarrow T(x, y) \leq T(x', y')$: monotonicity

iv) $T(T(x, y), z) = T(x, T(y, z))$: associativity □

Now we can easily recognize that the following operators hold conditions for t-norm.

(1) Intersection operator ($\cap$)

(2) Algebraic product operator (•)

(3) Bounded product operator (⊙)

(4) Drastic product operator (⋒)

Definition (t-conorm (s-norm))

$T : [0,1] \times [0,1] \rightarrow [0,1]$

$\forall x, y, x', y', z \in [0,1]$

i) $T(x, 0) = x, T(x, 1) = 1$: boundary condition

ii) $T(x, y) = T(y, x)$: commutativity

iii) $(x \leq x', y \leq y') \rightarrow T(x, y) \leq T(x', y')$: monotonicity

iv) $T(T(x, y), z) = T(x, T(y, z))$: associativity □

There are examples of t-conorm operators

(1) Union operator ($\cup$)

(2) Algebraic sum operator ($\hat{+}$)

(3) Bounded sum operator ($\oplus$)

(4) Drastic sum operator (⋓)

(5) Disjoint sum operator (Δ)

When computing the t-norm and t-conorm, operands in most cases are values of membership functions $\mu_A(x)$. An alternative symbol for these functions is *. And in other cases symbolTis used for t-norm, while symbol $\perp$ for t-conorm,

xTy : t-norm, $x \perp y$: t-conorm

t-norm is sometimes called s-norm. In practical applications, sometimes we could not determine which operator is appropriate to the application. In that case, we put the symbol * as a tentative operator. When we know that a t-norm operator can be used without selecting a specific operator, we put the symbolT. In this manner we can refine a system step by step. All t-norm and t-conorm functions follow these relations.

$$T(a, b) \leq Min[a, b]$$
$$\bot(a, b) \leq Max[a, b]$$

Let's see some operators following the above properties.

(1) $\wedge$: minimum

Instead of *, if $\wedge$ is applied

$$x \wedge 1 = x$$

Since this operator meets the previous conditions, it is a t-norm.

(2) $\vee$: maximum

If $\vee$ is applied instead of *,

$$x \vee 0 = x$$

then this becomes a t-conorm.

2.6.2 Duality of t-norms and t-conorms

We can see that duality exists between t-norm and t-conorm. Let function Trepresent a t-norm operator. If we define T's as

$$T'(x, y) = 1 - T(1-x, 1-y)$$

it becomes a t-conorm. That is , $\bot(x, y) = 1 - T(x, y)$

and again for $x, y \in [0,1]$, presume complements of x and y as

$$\bar{x} = 1 - x$$
$$\bar{y} = 1 - y$$

and complement of result from operation as,

$$\overline{x \top y} = 1 \mp (x, y)$$

then following relations are held, and they can be apprehended by De Morgan's law.

$$\bar{x} \bot \bar{y} = \overline{x \top y}$$
$$\bar{x} \top \bar{y} = \overline{x \bot y}$$

[SUMMARY]

☐ Standard operation of fuzzy set
  – Complement set : $\overline{A}$
    $$\mu_{\overline{A}}(x) = 1 - \mu_A(x)$$
  – Union set : $A \cup B$
    $$\mu_{A \cup B}(x) = Max[\mu_A(x), \mu_B(x)]$$
  – Intersection set : $A \cap B$
    $$\mu_{A \cap B}(x) = Min[\mu_A(x), \mu_B(x)]$$

☐ Difference of operation of crisp set and fuzzy set
  – No law of contradiction in fuzzy set
    $$A \cap \overline{A} \neq \varnothing$$
  – No law of exclude middle in fuzzy set
    $$A \cup \overline{A} \neq X$$

☐ Restriction for complement function
  – (C1) boundary condition
  – (C2) monotonic nonincreasing
  – (C3) continuous
  – (C4) involutive

☐ Example of fuzzy complement set
  – Yager complement set
    $$C_w(a) = (1 - a^w)^{1/w}, \quad w \in (-1, \infty)$$

☐ Condition for fuzzy union fuction
  – (U1) boundary condition
  – (U2) commutativity
  – (U3) monotonic function
  – (U4) associativity
  – (U5) continuous

☐ Example of fuzzy union function
  – Yager union function
    $$U_w(a, b) = Min[1, (a^w + b^w)^{1/w}], \quad w \in (0, \infty)$$

☐ Other union operators
   – Probabilistic sum of algebraic sum

   $$\forall x \in X, \quad \mu_{A\dot{+}B}(x) = \mu_A(x) + \mu_B(x) - \mu_A(x)\mu_B(x)$$

   – Bounded sum or bold union

   $$\forall x \in X, \quad \mu_{A\oplus B}(x) = Min[1, \mu_A(x) + \mu_B(x)]$$

   – Drastic sum

   $$\forall x \in X, \quad \mu_{A \,\overset{\vee}{\cdot}\, B}(x) = \begin{cases} \mu_A(x), & when \ \mu_B(x) = 0 \\ \mu_B(x), & when \ \mu_A(x) = 0 \\ 1, & others \end{cases}$$

   – Hamacher union function

   $$\forall x \in X, \quad \mu_{A\cup B}(x) = \frac{\mu_A(x) + \mu_B(x) - (2 - \gamma)\mu_A(x)\mu_B(x)}{1 - (1 - \gamma)\mu_A(x)\mu_B(x)}, \quad \gamma \geq 0$$

☐ Conditions for fuzzy intersection function
   – Boundary condition
   – Commutativity
   – Monotonic function
   – Associativity
   – Continuous

☐ Example of intersection function
   – Yager's intersection

   $$I_w(a,b) = 1 - Min[1, ((1 - a)^w + (1 - b)^w)^{1/w}] \quad where \ w \in (0, \infty)$$

☐ Other intersection operators
   – Algebraic product or probabilistic product

   $$\forall x \in X, \quad \mu_{A\bullet B}(x) = \mu_A(x) \bullet \mu_B(x)$$

   – Bounded product

   $$\forall x \in X, \quad \mu_{A \odot B}(x) = Max[0, \mu_A(x) + \mu_B(x) - 1]$$

   – Drastic product

   $$\forall x \in X, \quad \mu_{A \,\overset{\wedge}{\cdot}\, B}(x) = \begin{cases} \mu_A(x), & when \ \mu_B(x) = 1 \\ \mu_B(x), & when \ \mu_A(x) = 1 \\ 0, & when \ \mu_A(x), \mu_A(x) < 1 \end{cases}$$

   – Hamacher intersection

$$\forall x \in X, \quad \mu_{A \cap B}(x) = \frac{\mu_A(x)\mu_B(x)}{\gamma + (1-\gamma)(\mu_A(x) + \mu_B(x) - \mu_A(x)\mu_B(x))}, \quad \gamma \geq 0$$

☐ Other operations of fuzzy set
  − Disjunctive sum
$$A \oplus B = (A \cap \overline{B}) \cup (\overline{A} \cap B)$$
  Simple disjunctive sum and disjoint sum
  − Difference
    simple difference
$$A - B = A \cap \overline{B}$$
    bounded difference
$$\mu_{A \theta B}(x) = Max[0, \mu_A(x) - \mu_B(x)]$$

☐ Distance of fuzzy set
  − Hamming distance
$$d(A, B) = \sum_{\substack{i=1 \\ x_i \in X}}^{n} |\mu_A(x_i) - \mu_B(x_i)|$$
  relative hamming distance
$$\delta(A, B) = \frac{1}{n} d(A, B)$$
  symmetrical difference
$$\forall x \in X, \quad \mu_{A \nabla B}(x) = |\mu_A(x) - \mu_B(x)|$$

  − Euclidean distance
$$e(A, B) = \sqrt{\sum_{i=1}^{n} (\mu_A(x) - \mu_B(x))^2}$$
  − Minkowsk distance
$$d_w(A, B) = (\sum_{x \in X} |\mu_A(x) - \mu_B(x)|^w)^{1/w}, \quad w \in [1, \infty]$$

☐ Product of fuzzy set
  − Second power of fuzzy set : $A^2 = A \times A$
$$\mu_{A^2}(x) = [\mu_A(x)]^2, \quad \forall x \in X$$
  − Cartesian prduct : $A_1 \times A_2 \times \cdots \times A_n$
$$\mu_{A_1 \times A_2 \times \cdots \times A_n}(x_1, x_2, \cdots, x_n) = Min[\mu_{A_1}(x), \mu_{A_2}(x), \cdots, \mu_{A_n}(x)]$$

☐ t-norms operator
   – Intersection product ($\cap$)
   – Algebraic product ($\bullet$)
   – Bounded product ($\odot$)
   – Drastic product ($\widehat{\cap}$)

☐ t-conorms(s-norms) operator
   – Union ($\cup$)
   – Algebraic sum ($\hat{+}$)
   – Bounded sum ($\oplus$)
   – Drastic sum ($\underset{\smile}{}$)
   – Disjoint sum (Δ)

[EXERCISES]

2.1 Let sets A, B, and C be fuzzy sets defined on real numbers by the membership functions

$$\mu_A(x) = \frac{x}{x+1}, \ \mu_B(x) = \frac{1}{x^2+10}, \ \mu_C(x) = \frac{1}{10^x}$$

Determine mathematical membership functions and graphs of each of the followings :

a) $A \cup B$, $B \cap C$,

b) $A \cup B \cup C$, $A \cap B \cap C$

c) $A \cap \overline{C}$, $\overline{B} \cup C$

d) $\overline{A \cap B}$, $\overline{A \cup B}$

2.2 Show the two fuzzy sets satisfy the De Morgan's Law.

$$\mu_A(x) = \frac{1}{1+(x-10)}$$

$$\mu_B(x) = \frac{1}{1+x^2}$$

2.3 Show that the following sets don't satisfy the law of contradiction and the law of excluded middle.

a) $\mu_A(x) = \frac{1}{1+x}$

b) $A = \{(a, \ 0.4), (b, \ 0.5), (c, \ 0.9), (d, \ 1)\}$

2.4 Determine complements, unions, and intersections of the following sets by using Yager's operators for $\omega = 1, 2$

a) $A = \{(a, \ 0.5), (b, \ 0.9), (c, \ 0.1), (d, \ 0.5)\}$

b) $\mu_A(x) = \frac{1}{1+x}$

2.5 Compute the complements of the following sets by Yager's complements $w = 1, 2$.

a) $A = \{(a, \ 0.5), (b, \ 0.9), (c, \ 0.3)\}$

b) $\mu_A(x) = \frac{1}{1+(x-1)}$

2.6 Compute the complements of the following sets by using Probabilistic, Bounded, Drastic, and Hamacher product.

a) $\mu_A(x) = \dfrac{1}{1+x^2}$

b) $\mu_A(x) = 2^{-x}$

c) $A = \{(a,\ 0.4),\ (b,\ 0.5),\ (c,\ 0.9)\}$

2.7 Compute the simple disjunctive sum, disjoint sum, simple difference, and bounded difference of the sets

$$A = \{(x,\ 0.5),\ (y,\ 0.4),\ (z,\ 0.9),\ (w,\ 0.1)\}$$
$$B = \{(x,\ 0.4),\ (y,\ 0.8),\ (z,\ 0.1),\ (w,\ 1)\}$$

2.8 Determine the distances (Hamming, Euclidean and Minkowski for $\omega = 2$) between the following sets

$$A = \{(x,\ 0.5),\ (y,\ 0.4),\ (z,\ 0.9),\ (w,\ 0.1)\}$$
$$B = \{(x,\ 0.1),\ (y,\ 0.9),\ (z,\ 0.1),\ (w,\ 0.9)\}$$

2.9 Prove the following properties.

a) Let function Ta t-norm operation.

The following T′is a t-conorm operation.

$$T'\ (x,\ y) = 1 - T'(1\text{-}x,\ 1\text{-}y)$$

b) $\overline{x \perp y} = \overline{x}\ T\ \overline{y}$ and $\overline{x} T \overline{y} = \overline{x \perp y}$

where $\overline{x} = 1 - x$, $\overline{y} = 1 - y$ and $\overline{x T y} = 1 \mp\ (x, y)$

where $\perp$ represents t-conorm operator.

2.10 Determine the closet pair of sets among the following sets

$A = \{(x_1,\ 0.4),\ (x_2,\ 0.4),\ (x_3,\ 0.9),\ (x_4,\ 0.5)\}$
$B = \{(x_1,\ 0.1),\ (x_2,\ 0.0),\ (x_3,\ 0.9)\}$
$C = \{(x_1,\ 0.5),\ (x_2,\ 0.5),\ (x_3,\ 0.9)\}$

2.11 Show the Max operator satisfies the properties boundary condition, commutativity, associativity, continuity, and idempotency.

2.12 Prove the following equation.

$$0 \le \mu_{A \oplus B}(x) \le 0.5$$

where $A \oplus B = (A \cap \overline{B}) \cup (\overline{A} \cap B)$

that is, $\oplus$ is simple disjunctive sum operator.

Chapter 3. FUZZY RELATION AND COMPOSITION

The concept of fuzzy set as a generalization of crisp set has been introduced in the previous chapter. Relations between elements of crisp sets can be extended to fuzzy relations, and the relations will be considered as fuzzy sets. In this chapter, we should be familiar with the proper meanings of the two terms: ***crisp relation*** and ***fuzzy relation***. Various operations on the fuzzy relations will be introduced.

3.1 Crisp Relation

3.1.1 Product Set

Assume that an order between elements x and y exists, the ***pair*** made of these two elements is called an ***ordered pair***. An ordered pair is usually denoted by (x, y).

Definition (Product set) Let A and B be two non-empty sets, the product set or Cartesian product $A \times B$ is defined as follows,

$$A \times B = \{(a, b) \mid a \in A, b \in B \}$$

The concept of Cartesian product can be extended to n sets. For an arbitrary number of sets $A_1, A_2, \ldots, A_n$, the set of all n-tuples $(a_1, \ldots, a_n)$ such that $a_1 \in A_1, a_2 \in A_2, \ldots, a_n \in A_n$, is called the Cartesian product and is written as $A_1 \times A_2 \times \ldots \times A_n$ or

$$\prod_{i=1}^{n} A_i \quad \square$$

Instead of $A \times A$ and $A \times A \times A \times \ldots \times A$, we use the notations A^2 and A^n respectively. The product is used for the "composition" of sets and relations in the later sections. For example, a relation is a product space obtained from two sets A and B. $\mathcal{R}^3 = \mathcal{R} \times \mathcal{R} \times \mathcal{R}$ denotes the 3-dimensional space of real numbers.

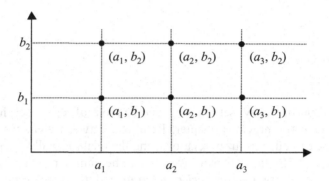

Fig. 3.1. Product set $A \times B$

Example 3.1 When $A = \{a_1, a_2, a_3\}$, $B = \{b_1, b_2\}$ the Cartesian product yields (Fig 3.1)

$$A \times B = \{(a_1, b_1), (a_1, b_2), (a_2, b_1), (a_2, b_2), (a_3, b_1), (a_3, b_2)\}$$

The Cartesian product $A \times A$ is in the following and is also shown in (Fig 3.2)

$$A \times A = \{(a_1, a_1), (a_1, a_2), (a_1, a_3), (a_2, a_1), (a_2, a_2), (a_2, a_3),$$
$$(a_3, a_1), (a_3, a_2), (a_3, a_3)\} \quad \square$$

3.1.2 Definition of Relation

Definition (Binary Relation) If A and B are two sets and there is a specific property between elements x of A and y of B, this property can be described using the ordered pair (x, y). A set of such (x, y) pairs, $x \in A$ and $y \in B$, is called a relation R.

$$R = \{ (x,y) \mid x \in A, y \in B \}$$

R is a binary relation and a subset of $A \times B$. $\quad \square$

The term "x is in relation R with y" is denoted as
$$(x, y) \in R \text{ or } x \, R \, y \text{ with } R \subseteq A \times B.$$

If $(x, y) \notin R$, x is not in relation R with y. If $A = B$ or R is a relation from A to A, it is written
$$(x, x) \in R \text{ or } x \, R \, x \text{ for } R \subseteq A \times A.$$

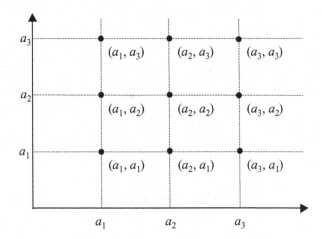

Fig. 3.2. Cartesian product $A \times A$

Definition (n-ary relation) For sets A_1, A_2, A_3, ..., A_n, the relation among elements $x_1 \in A_1$, $x_2 \in A_2$, $x_3 \in A_3$, ..., $x_n \in A_n$ can be described by n-tuple $(x_1, x_2, ..., x_n)$. A collection of such n-tuples $(x_1, x_2, x_3, ..., x_n)$ is a relation R among A_1, A_2, A_3, ..., A_n. That is

$$(x_1, x_2, x_3, \dots, x_n) \in R$$
$$R \subseteq A_1 \times A_2 \times A_3 \times \dots \times A_n \quad \square$$

Definition (Domain and range) Let R stand for a relation between A and B. The domain and range of this relation are defined as follows (Fig. 3.3) :

$$dom(R) = \{ x \mid x \in A, (x, y) \in R \text{ for some } y \in B \}$$
$$ran(R) = \{ y \mid y \in B, (x, y) \in R \text{ for some } x \in A \} \quad \square$$

Here we call set A as support of $dom(R)$ and B as support of $ran(R)$. $dom(R) = A$ results in completely specified and $dom(R) \subseteq A$ incompletely specified. The relation $R \subseteq A \times B$ is a set of ordered pairs (x, y). Thus, if we have a certain element x in A, we can find y of B, i.e., the mapped image of A. We say "y is the mapping of x" (Fig 3.4).

If we express this mapping as f, y is called the **image** of x which is denoted as $f(x)$

$$R = \{(x, y) \mid x \in A, y \in B, y = f(x)\} \text{ or } f : A \to B$$

So we might say $ran(R)$ is the set gathering of these $f(x)$

$$ran(R) = f(A) = \{f(x) \mid x \in A\}$$

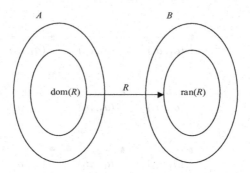

Fig. 3.3. Domain and range

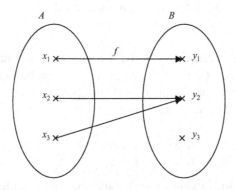

Fig. 3.4. Mapping $y = f(x)$

3.1.3 Characteristics of Relation

We can summarize the properties of relation as follows :

(1) One-to-many *(Fig. 3.5)*
     R is said to be one-to-many if
$$\exists x \in A, \quad y_1, \quad y_2 \in B \quad (x, y_1) \in R, (x, y_2) \in R$$

(2) Surjection *(may-to-one)*
     R is said to be a surjection if $f(A) = B$ or $ran(R) = B$.
$$\forall y \in B, \exists x \in A, y = f(x)$$

Thus, even if x1 ≠ x2, f(x1) = f(x2) can hold.

(3) Injection (into, one-to-one)
     R is said to be an injection if for all x_1, $x_2 \in A$,
               if $x_1 \neq x_2$, $f(x_1) \neq f(x_2)$.
     Therefore, if R is an injection, $(x_1, y) \in R$ and $(x_2, y) \in R$ then $x_1 = x_2$.

(4) Bijection (one-to-one correspondence)
     R is said to be a bijection if it is both a surjection and an injection.
     Assuming A and B are in bijection, this is an equivalence relation
     holds, i.e. the elements and the number of the elements correspond.

   If each member of the domain appears exactly once in *R*, the relation *R*
is called a mapping or a function. When at least one member of the domain
is related to more than one element of the range, the relation is not a
mapping and is instead called one-to-many relation. Therefore surjection,
injection and bijection are functions, and thus an element *x* in *dom(R)* is
mapped to only one element *y* in *ran(R)* by them.

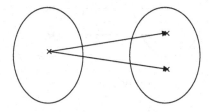

Fig. 3.5. One-to-many relation (*not a function*)

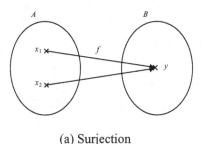

(a) Surjection

Fig. 3.6. Functions (surjection, injection and bijection)

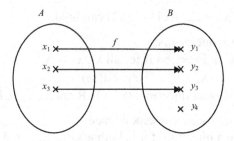

(b) Injection

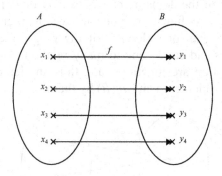

(c) Bijection

Fig. 3.6. (cont')

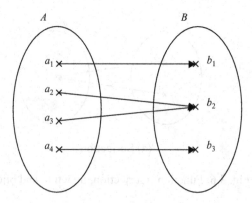

Fig. 3.7. Binary relation from A to B

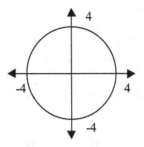

Fig. 3.8. Relation of $x^2 + y^2 = 4$

3.1.4 Representation Methods of Relations

There are four methods of expressing the relation between sets A and B.

(1) Bipartigraph
     The first is by illustrating A and B in a figure and representing the relation by drawing arcs or edges (Fig 3.7).

(2) Coordinate diagram
     The second is to use a coordinate diagram by plotting members of A on x axis and that of B on y axis, and then the members of $A \times B$ lie on the space. Fig 3.8 shows this type of representation for the relation R, namely $x^2 + y^2 = 4$ where $x \in A$ and $y \in B$.

(3) Matrix
     The third method is by manipulating relation matrix. Let A and B be finite sets having m and n elements respectively. Assuming R is a relation between A and B, we may represent the relation by matrix $M_R = (m_{ij})$ which is defined as follows

$$M_R = (m_{ij})$$

$$m_{ij} = \begin{cases} 1, & (a_i, b_j) \in R \\ 0, & (a_i, b_j) \notin R \end{cases}$$

$$i = 1, 2, 3, \ldots, m$$
$$j = 1, 2, 3, \ldots, n$$

Such matrix is called a relation matrix, and that of the relation in (Fig. 3.7) is given in the following.

R	b_1	b_2	b_3
a_1	1	0	0
a_2	0	1	0
a_3	0	1	0
a_4	0	0	1

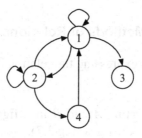

Fig. 3.9. Directed graph

(4)   Digraph

The fourth method is the directed graph or digraph method. Elements are represented as nodes, and relations between elements as directed edges.

$A = \{1, 2, 3, 4\}$ and $R = \{(1, 1), (1, 2), (2, 1), (2, 2), (1, 3), (2, 4), (4, 1)\}$ for instance. (Fig 3.9) shows the directed graph corresponding to this relation. When a relation is symmetric, an undirected graph can be used instead of the directed graph.

3.1.5 Operations on Relations

In the previous section, relation R was defined as a set. That is, R is a set containing ordered pairs (x, y) for $x \in A$, $y \in B$. If we assume R and S are relations defined on the same space $A \times B$, these relations might have operations of union, intersection, inverse and composition.

(1)   Union of relation
      $T = R \cup S$ is said to be the union of R and S.
                  If $(x, y) \in R$ or $(x, y) \in S$, then $(x, y) \in T$

(2)   Intersection of relation
      $T = R \cap S$ is said to be the intersection of R and S.
                  If $(x, y) \in R$ and $(x, y) \in S$, then $(x, y) \in T$.

(3)   Complement of relation
      $A \times B$ represents all possible relations that can occur between two sets.
      That means it is equivalent to the concept of universal set. Now

$$\overline{R} = \{(x, y) \in A \times B \,|\, (x, y) \notin R\}$$

That is,
if $(x, y) \notin R$, then $(x, y) \in \overline{R}$

      $\overline{R}$ is said to be the complement of the relation R.

(4)   Inverse relation
      Let R be a relation from A to B. The inverse R^{-1} is defined as,
              $R^{-1} = \{(y, x) \in B \times A \,|\, (x, y) \in R, x \in A, y \in B\}$

(5)   Composition
      Let R and S be two relations defined on sets A, B and C. T is said to be
      a composition of R and S.
          $R \subseteq A \times B, S \subseteq B \times C$
          $T = S \bullet R \subseteq A \times C$
          $T = \{(x, z) \,|\, x \in A, y \in B, z \in C, (x, y) \in R, (y, z) \in S\}$
      Let R be a relation characterizing the set A. The composition of R and
      R is written as $R \bullet R$ or R^2. R^n is the n-th composition of R.

3.1.6 Path and Connectivity in Graph

Path of length n in the graph defined by a relation $R \subseteq A \times A$ is a finite
series of $p = a, x_1, x_2, ..., x_{n-1}$, b where each element should be $a\ R\ x_1$, $x_1\ R$
$x_2, ..., x_{n-1}\ R\ b$. Besides, when n refers to a positive integer,

(1)   Relation R^n on A is defined, $x\ R^n\ y$ means there exists a path from x to
      y whose length is n.

(2)   Relation R^∞ on A is defined, $x\ R^\infty\ y$ means there exists a path from x
      to y. That is, there exists x
      $R\ y$ or $x\ R^2\ y$ or $x\ R^3\ y$ ...   and. This relation R^∞ is the reachability
      relation, and denoted as $xR^\infty y$

(3) The reachability relation R^∞ can be interpreted as connectivity relation of A.

3.2 Properties of Relation on A Single Set

Now we shall see the fundamental properties of relation defined on a set, that is, $R \subseteq A \times A$. We will review the properties such as reflexive relation, symmetric relation, transitive relation, closure, equivalence relation, compatibility relation, pre-order relation and order relation in detail.

3.2.1 Fundamental Properties

(1)   Reflexive relation
     If for all $x \in A$, the relation xRx or $(x, x) \in R$ is established, we call it reflexive relation. The reflexive relation might be denoted as

$$x \in A \rightarrow (x, x) \in R \text{ or } \mu_R(x, x) = 1, \forall x \in A$$
where the symbol "$\rightarrow$" means "implication"
     If it is not satisfied for some $x \in A$, the relation is called "irreflexive". if it is not satisfied for all $x \in A$, the relation is "antireflexive".
     When you convert a reflexive relation into the corresponding relation matrix, you will easily notice that every diagonal member is set to 1. A reflexive relation is often denoted by D.

(2)   Symmetric relation
     For all $x, y \in A$, if $xRy = yRx$, R is said to be a symmetric relation and expressed as
$$(x, y) \in R \rightarrow (y, x) \in R \text{ or}$$
$$\mu_R(x, y) = \mu_R(y, x), \forall x, y \in A$$
     The relation is "asymmetric" or "nonsymmetric" when for some $x, y \in A$, $(x, y) \in R$ and $(y, x) \notin R$. It is an "antisymmetric" relation if for all $x, y \in A$, $(x, y) \in R$ and $(y, x) \notin R$

(3)   Transitive relation
     This concept is achieved when a relation defined on A verifies the following property.
     For all $x, y, z \in A$
$$(x, y) \in R, (y, z) \in R \rightarrow (x, z) \in R$$

(4)   Closure
     When relation R is defined in A, the requisites for closure are,

1) Set A should satisfy a certain specific property.
2) Intersection between A's subsets should satisfy the relation R.

The smallest relation R' containing the specific property is called closure of R.

Example 3.2 If R is defined on A, assuming R is not a reflexive relation, then $R' = D \cup R$ contains R and reflexive relation. At this time, R' is said to be the reflexive closure of R.   □

Example 3.3 If R is defined on A, transitive closure of R is as follows(Fig 3.10), which is the same as R^{∞} (reachability relation).

$$R^{\infty} = R \cup R^2 \cup R^3 \cup \ldots$$

The transitive closure R^{∞} of R for $A = \{1, 2, 3, 4\}$ and $R = \{(1, 2), (2, 3), (3, 4), (2, 1)\}$ is,
$R^{\infty} = \{(1, 1), (1, 2), (1, 3), (1,4), (2,1), (2,2), (2, 3), (2, 4), (3, 4)\}$.
(Fig 3.10) explains this example □

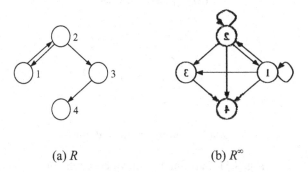

(a) R (b) R^{∞}

Fig. 3.10. Transitive closure

3.2.2 Equivalence Relation

Definition (Equivalence relation) Relation $R \subseteq A \times A$ is an equivalence relation if the following conditions are satisfied.

i).   Reflexive relation
$$x \in A \rightarrow (x, x) \in R$$
ii).   Symmetric relation
$$(x, y) \in R \rightarrow (y, x) \in R$$
iii).   Transitive relation
$$(x, y) \in R, (y, z) \in R \rightarrow (x, z) \in R   □$$

If an equivalence relation R is applied to a set A, we can perform a partition of A into n disjoint subsets A_1, A_2, ... which are equivalence classes of R. At this time in each equivalence class, the above three conditions are verified. Assuming equivalence relation R in A is given, equivalence classes are obtained. The set of these classes is a partition of A

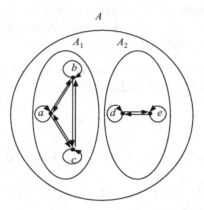

(a) Expression by set

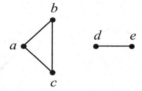

(b) Expression by undirected graph

Fig. 3.11. Partition by equivalence relation

by R and denoted as $\pi(A/R)$. Fig 3.11 shows the equivalence relation verified in A_1 and A_2.

$$\pi(A/R) = \{A_1, A_2\} = \{\{a, b, c\}, \{d, e\}\}$$

3.2.3 Compatibility Relation (Tolerance Relation)

Definition (Compatibility relation) If a relation satisfies the following conditions for every $x, y \in A$, the relation is called compatibility relation.

i)   Reflexive relation

$$x \in A \rightarrow (x, x) \in R$$

ii) Symmetric relation

$$(x, y) \in R \rightarrow (y, x) \in R \quad \square$$

If a compatibility relation R is applied to set A, we can decompose the set A into disjoint subsets which are compatibility classes. In each compatibility class, the above two conditions are satisfied. Therefore, a compatibility relation on a set A gives a partition. But the only difference from the equivalence relation is that transitive relation is not completed in the compatibility relation.

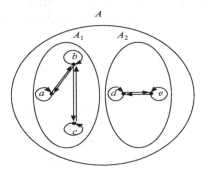

(a) Expression by set

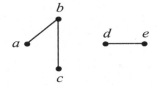

(b) Expression by undirected graph

Fig. 3.12. Partition by compatibility relation

(Fig.3.12) describes a partition of set A by a compatibility relation. Here, compatibility classes are $\{a, b, c\}$ and $\{d, e\}$.

3.2.4 Pre-order Relation

Definition (Pre-order relation) For any $x, y, z \in A$, if a relation $R \subseteq A \times A$ obeys these two conditions, it is called pre-order relation.
   i)   Reflexive relation
$$x \in A \rightarrow (x, x) \in R$$
   ii) Transitive relation
$$(x, y) \in R, (y, z) \in R \rightarrow (x, z) \in R \quad \square$$

Unlike for the equivalence relation, no symmetric relation exists here. That is, whether elements in A are symmetric or not is not our concern here. Fig 3.13 is an example of pre-order for given relation R on A. Edges (b, d) and (f, h) are symmetric and others are non-symmetric. Looking over (b, d) and (f, h), you shall notice that an equivalence relation holds. So these members can be tided up as the same classes as in the figure.

Consequently we can assure that if a pre-order exists, it implies that an order exists between classes, and that the number of members in a class can be more than 1. If the property of antisymmetric relation is added to the pre-order, the number of member in a class should be 1 and it becomes an order relation.

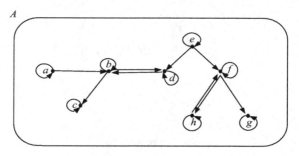

(a) Pre-order relation

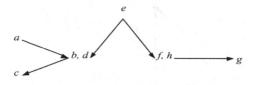

(b) Pre-order

Fig. 3.13. Pre-order relation

3.2.5 Order Relation

Definition (Order relation) If a binary relation $R \subseteq A \times A$ satisfies the followings for any $x, y, z \in A$, it is called order relation (or partial order relation).

i)   Reflexive relation
$$x \in A \to (x, x) \in R$$
ii) Antisymmetric relation
$$(x, y) \in R \to (y, x) \notin R$$
iii) Transitive relation
$$(x, y) \in R, (y, z) \in R \to (x, z) \in R \quad \square$$

When relation R is given to an arbitrary set A, an order according to R is defined among the elements of A (Fig. 3.14). If the condition (1) is replaced by

(i') Antireflexive relation

$$x \in A \rightarrow (x, x) \notin R$$

we apply the term "strict order relation" for it.

Since the order is determined in (b, a) and (d, g) in the figure, they can be compared. In this case we say "comparable". But a and c are "incomparable". Comparable pair (x, y) is denoted as $x \geq y$.

In the order relation, when the following condition (4) is added, we call this relation a "total order" or "linear order" relation.

iv) $\forall x, y \in A, (x, y) \in R$ or $(y, x) \in R$

The total order is also termed as a "chain" since it can be drawn in a line. Comparing to the total order, the order following only conditions (1) ,(2) and (3) is called a "partial order", and a set defining the partial order is called "partial order set".

Definition (Ordinal function) For all $x, y \in A$ $(x \neq y)$,

i)   If $(x, y) \in R$, xRy or $x > y$,

$$f(x) = f(y) + 1$$

ii) If reachability relation exists in x and y, i.e. if $xR^{\infty}y$,

$$f(x) > f(y) \qquad\qquad \square$$

The pair (x, y) can be written as $x \geq y$. Applying the ordinal function into the order relation in (Fig. 3.14(a)), yields (Fig. 3.14(b).)

Now we can summarize as follows :

(1) In the pre-order, the symmetry or nonsymmetry is allowed. But in the case of order, only the antisymmetry is allowed. In other words, adding the antisymmetry to the pre-order, we get an order.

(2) A pre-order is said to be an order between classes. In other words, an order is a pre-order restricting that the number of class is 1.

(3) An equivalence relation has symmetry, so it can be obtained by adding the symmetry to the pre-order relation.

Characteristics so far discussed are summaried in Table 3.4.

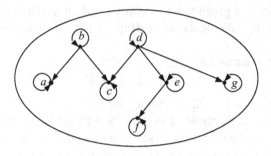

(a) Order relation

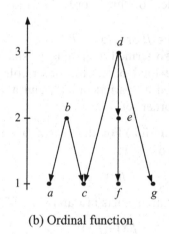

(b) Ordinal function

Fig. 3.14. Order relation and ordinal function

3.3 Fuzzy Relation

3.3.1 Definition of Fuzzy Relation

If a crisp relation R represents that of from sets A to B, for $x \in A$ and $y \in B$, its membership function $\mu_R(x, y)$ is,

$$\mu_R(x, y) = \begin{cases} 1 \text{ iff } (x, y) \in R \\ 0 \text{ iff } (x, y) \notin R \end{cases}$$

This membership function maps $A \times B$ to set $\{0, 1\}$.

$$\mu_R : A \times B \rightarrow \{0, 1\}$$

Table 3.1 Comparison of relations

Property / Relation	Reflexive	Antireflexive	Symmetric	Antisymmetric	Transitive
Equivalence	✓		✓		✓
Compatibility	✓		✓		
Pre-order	✓				✓
Order	✓			✓	✓
Strict order		✓		✓	✓

This membership function maps $A \times B$ to set $\{0, 1\}$.

$$\mu_R : A \times B \to \{0, 1\}$$

We know that the relation R is considered as a set. Recalling the previous fuzzy concept, we can define ambiguous relation.

Definiton (Fuzzy relation) Fuzzy relation has degree of membership whose value lies in $[0, 1]$.

$$\mu_R : A \times B \to [0, 1]$$

$$R = \{((x, y), \mu_R(x, y)) | \mu_R(x, y) \geq 0 , \quad x \in A, \quad y \in B\} \quad \square$$

Here $\mu_R(x, y)$ is interpreted as strength of relation between x and y. When $\mu_R(x, y) \geq \mu_R(x', y')$, (x, y) is more strongly related than (x', y').

When a fuzzy relation $R \subseteq A \times B$ is given, this relation R can be thought as a fuzzy set in the space $A \times B$. Fig 3.15 illustrates that the fuzzy relation R is a fuzzy set of pairs (a, b) of elements where $a_i \in A$, $b_i \in B$.

Let's assume a Cartesian product space $X_1 \times X_2$ composed of two sets X_1 and X_2. This space makes a set of pairs (x_1, x_2) for all $x_1 \in X_1$, $x_2 \in X_2$. Given a fuzzy relation R between two sets X_1 and X_2, this relation is a set of pairs $(x_1, x_2) \in R$. Consequently, this fuzzy relation can be presumed to be a fuzzy restriction to the set $X_1 \times X_2$. Therefore, $R \subseteq X_1 \times X_2$

Fuzzy binary relation can be extended to n-ary relation. If we assume $X_1, X_2, ..., X_n$ to be fuzzy sets, fuzzy relation $R \subseteq X_1 \times X_2 \times ... \times X_n$ can be said to be a fuzzy set of tuple elements $(x_1, x_2, ..., x_n)$, where $x_1 \in X_1$, $x_2 \in X_2, ..., x_n \in X_n$.

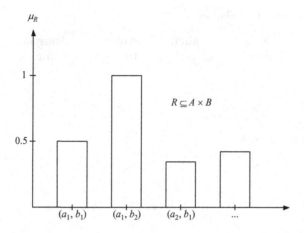

Fig. 3.15. Fuzzy relation as a fuzzy set A×B

3.3.2 Examples of Fuzzy Relation

Example 3.3 (Fig 3.16) for instance, crisp relation R in the figure (a) reflects a relation in $A \times A$. Expressing this by membership function, $\mu_R(a, c) = 1$, $\mu_R(b, a) = 1$, $\mu_R(c, b) = 1$ and $\mu_R(c, d) = 1$.

If this relation is given as the value between 0 and 1 as in the figure (b), this relation becomes a fuzzy relation. Expressing this fuzzy relation by membership function yields,

$$\mu_R(a, c) = 0.8, \ \mu_R(b, a) = 1.0, \ \mu_R(c, b) = 0.9, \ \mu_R(c, d) = 1.0$$

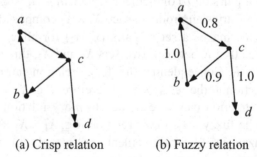

(a) Crisp relation   (b) Fuzzy relation

Fig. 3.16. Crisp and Fuzzy relations

It's corresponding fuzzy matrix is as follows.

A	a	b	c	d
a	0.0	0.0	0.8	0.0
b	1.0	0.0	0.0	0.0
c	0.0	0.9	0.0	1.0
d	0.0	0.0	0.0	0.0

Fuzzy relation is mainly useful when expressing knowledge. Generally, the knowledge is composed of rules and facts. A rule can contain the concept of possibility of event b after event a has occurred. For instance, let us assume that set A is a set of events and R is a rule. Then by the rule R, the possibility for the occurrence of event c after event a occurred is 0.8 in the previous fuzzy relation.

When crisp relation R represents the relation from crisp sets A to B, its domain and range can be defined as,

$$dom(R) = \{x \mid x \in A, y \in A, \mu_R(x, y) = 1\}$$

$$ran(R) = \{y \mid x \in A, y \in A, \mu_R(x, y) = 1\}$$

Definition (Domain and range of fuzzy relation) When fuzzy relation R is defined in crisp sets A and B, the domain and range of this relation are defined as :

$$\mu_{dom(R)}(x) = \max_{y \in B} \mu_R(x, y)$$

$$\mu_{ran(R)}(y) = \max_{x \in A} \mu_R(x, y)$$

Set A becomes the support of $dom(R)$ and $dom(R) \subseteq A$. Set B is the support of $ran(R)$ and

$$ran(R) \subseteq B. \quad \square$$

3.3.3 Fuzzy Matrix

Given a certain vector, if an element of this vector has its value between 0 and 1, we call this vector a fuzzy vector. Fuzzy matrix is a gathering of such vectors. Given a fuzzy matrix $A = (a_{ij})$ and $B = (b_{ij})$, we can perform operations on these fuzzy matrices.

(1) Sum
$$A + B = \text{Max}\ [a_{ij},\ b_{ij}]$$

(2) Max product
$$A \bullet B = AB = \text{Max}_{k}\ [\ \text{Min}\ (a_{ik},\ b_{kj})\ ]$$

(3) Scalar product
$$\lambda A \quad \text{where } 0 \le \lambda \le 1$$

Example 3.4 The followings are examples of sum and max product on fuzzy sets A and B.

$$A = \quad
\begin{array}{c|ccc}
  & a & b & c \\
\hline
a & 0.2 & 0.5 & 0.0 \\
b & 0.4 & 1.0 & 0.1 \\
c & 0.0 & 1.0 & 0.0 \\
\end{array}
\qquad
B = \quad
\begin{array}{c|ccc}
  & a & b & c \\
\hline
a & 1.0 & 0.1 & 0.0 \\
b & 0.0 & 0.0 & 0.5 \\
c & 0.0 & 1.0 & 0.1 \\
\end{array}$$

$$A + B = \quad
\begin{array}{c|ccc}
  & a & b & c \\
\hline
a & 1.0 & 0.5 & 0.0 \\
b & 0.4 & 1.0 & 0.5 \\
c & 0.0 & 1.0 & 0.1 \\
\end{array}
\qquad
A \bullet B = \quad
\begin{array}{c|ccc}
  & a & b & c \\
\hline
a & 0.2 & 0.1 & 0.5 \\
b & 0.4 & 0.1 & 0.5 \\
c & 0.0 & 0.0 & 0.5 \\
\end{array}$$

Here let's have a closer look at the product $A \bullet B$ of A and B. For instance, in the first row and second column of the matrix $C = A \bullet B$, the value 0.1 ($C_{12} = 0.1$) is calculated by applying the Max-Min operation to the values of the first row (0.2, 0.5 and 0.0) of A, and those of the second column (0.1 , 0.0 and 1.0) of B.

$$
\begin{array}{ccc}
0.2 & 0.5 & 0.0 \\
0.1 & 0.0 & 1.0 \\
\end{array}
$$

$Min \Downarrow$ _____

$$
\begin{array}{ccccc}
0.1 & 0.0 & 0.0 & \Rightarrow & 0.1 \\
 & & & Max & \\
\end{array}
$$

In the same manner $C_{13} = 0.5$ is obtained by applying the same procedure of calculation to the first row (0.2, 0.5, 0.0) of A and the third column of B (0.0, 0.5, 0.1).

$$\begin{array}{ccc} 0.2 & 0.5 & 0.0 \\ 0.0 & 0.5 & 0.1 \end{array}$$

$Min \Downarrow$ _____

$$\begin{array}{ccc} 0.0 & 0.5 & 0.0 \end{array} \quad \Rightarrow \quad 0.5 \qquad \square$$

$$Max$$

And for all i and j, if $a_{ij} \le b_{ij}$ holds, matrix B is bigger than A.

$$a_{ij} \le b_{ij} \quad \Leftrightarrow \quad A \le B$$

Also when $A \le B$ for arbitrary fuzzy matrices S and T, the following relation holds from the Max-Product operation.

$$A \le B \quad \Leftrightarrow \quad SA \le SB, AT \le BT$$

Definition (Fuzzy relation matrix) If a fuzzy relation R is given in the form of fuzzy matrix, its elements represent the membership values of this relation. That is, if the matrix is denoted by M_R, and membership values by $\mu_R (i, j)$, then $\quad M_R = (\mu_R (i, j)) \qquad \square$

3.3.4 Operation of Fuzzy Relation

We know now a relation is one kind of sets. Therefore we can apply operations of fuzzy set to the relation. We assume $R \subseteq A \times B$ and $S \subseteq A \times B$.

(1) Union relation
      Union of two relations R and S is defined as follows :

$$\forall (x, y) \in A \times B$$
$$\mu_{R \cup S} (x, y) = Max [\mu_R (x, y), \mu_S (x, y)]$$
$$= \mu_R (x, y) \vee \mu_S (x, y)$$

We generally use the sign $\vee$ for Max operation. For n relations, we extend it to the following.

$$\mu_{R_1 \cup R_2 \cup R_3 \cup ... \cup R_n} (x, y) = \bigvee_{R_i} \mu_{R_i} (x, y)$$

If expressing the fuzzy relation by fuzzy matrices, i.e. M_R and M_S, matrix $M_{R \cup S}$ concerning the union is obtained from the sum of two matrices $M_R + M_S$.

$$M_{R \cup S} = M_R + M_S$$

(2) Intersection relation
      The intersection relation $R \cap S$ of set A and B is defined by the following membership function.

$$\mu_{R \cap S} (x) = \text{Min } [\mu_R (x, y), \mu_S (x, y)]$$
$$= \mu_R (x, y) \wedge \mu_S (x, y)$$

The symbol $\wedge$ is for the Min operation. In the same manner, the intersection relation for n relations is defined by

$$\mu_{R_1 \cap R_2 \cap R_3 \cap \dots \cap R_n} (x, y) = \bigwedge_{R_i} \mu_{Ri} (x, y)$$

(3) Complement relation

Complement relation $\overline{R}$ for fuzzy relation R shall be defined by the following membership function.

$$\forall (x, y) \in A \times B \quad \mu_{\overline{R}}(x, y) = 1 - \mu_R (x, y)$$

Example 3.5 Two fuzzy relation matrices M_R and M_S are given.

M_R	a	b	c
1	0.3	0.2	1.0
2	0.8	1.0	1.0
3	0.0	1.0	0.0

M_S	a	b	c
1	0.3	0.0	0.1
2	0.1	0.8	1.0
3	0.6	0.9	0.3

Fuzzy relation matrices $M_{R \cup S}$ and $M_{R \cap S}$ corresponding $R \cup S$ and $R \cap S$ yield the followings.

$M_{R \cup S}$	a	b	c
1	0.3	0.2	1.0
2	0.8	1.0	1.0
3	0.6	1.0	0.3

$M_{R \cap S}$	a	b	c
1	0.3	0.0	0.1
2	0.1	0.8	1.0
3	0.0	0.9	0.0

Also complement relation of fuzzy relation R shall be

$M_{\overline{R}}$	a	b	c
1	0.7	0.8	0.0
2	0.2	0.0	0.0
3	1.0	0.0	1.0

(4) Inverse relation

When a fuzzy relation $R \subseteq A \times B$ is given, the inverse relation of R^{-1} is defined by the following membership function.

For all $(x, y) \subseteq A \times B$, $\mu_R^{-1} (y, x) = \mu_R (x, y)$

3.3.5 Composition of Fuzzy Relation

Definition (Composition of fuzzy relation) Two fuzzy relations R and S are defined on sets A, B and C. That is, $R \subseteq A \times B$, $S \subseteq B \times C$. The composition $S \bullet R = SR$ of two relations R and S is expressed by the relation from A to C, and this composition is defined by the following.
For $(x, y) \in A \times B$, $\quad (y, z) \in B \times C$,

$$\mu_{S \bullet R}(x, z) = \underset{y}{\text{Max}}[\text{Min}(\mu_R(x, y), \mu_S(y, z))]$$

$$= \underset{y}{\vee}[\mu_R(x, y) \wedge \mu_S(y, z)] \qquad \square$$

$S \bullet R$ from this elaboration is a subset of $A \times C$. That is, $S \bullet R \subseteq A \times C$.

If the relations R and S are represented by matrices M_R and M_S, the matrix $M_{S \bullet R}$ corresponding to $S \bullet R$ is obtained from the product of M_R and M_S.

$$M_{S \bullet R} = M_R \bullet M_S$$

Example 3.6 Consider fuzzy relations $R \subseteq A \times B$, $S \subseteq B \times C$. The sets A, B and C shall be the sets of events. By the relation R, we can see the possibility of occurrence of B after A, and by S, that of C after B. For example, by M_R, the possibility of $a \in B$ after $1 \in A$ is 0.1. By M_S, the possibility of occurrence of α after a is 0.9.

R	a	b	c	d
1	0.1	0.2	0.0	1.0
2	0.3	0.3	0.0	0.2
3	0.8	0.9	1.0	0.4

S	α	β	γ
a	0.9	0.0	0.3
b	0.2	1.0	0.8
c	0.8	0.0	0.7
d	0.4	0.2	0.3

Here, we can not guess the possibility of C when A is occurred. So our main job now will be the obtaining the composition $S \bullet R \subseteq A \times C$. The following matrix $M_{S \bullet R}$ represents this composition and it is also given in (Fig 3.17).

Now we see the possibility of occurrence of $\alpha \in C$ after event $1 \in A$ is 0.4, and that for $\beta \in C$ after event $2 \in A$ is 0.3.etc... $\square$

Presuming that the relations R and S are the expressions of rules that guide the occurrence of event or fact. Then the possibility of occurrence of event B when event A is happened is guided by the rule R. And rule S indicates the possibility of C when B is existing. For further cases, the possibility of C when A has occurred can be induced from the composition rule $S \bullet R$. This manner is named as an "inference" which is a process producing new information.

S•R	α	β	γ
1	0.4	0.2	0.3
2	0.3	0.3	0.3
3	0.8	0.9	0.8

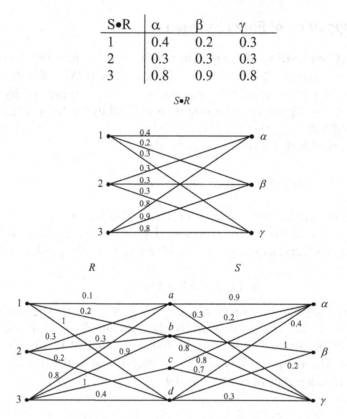

Fig. 3.17. Composition of fuzzy relation

3.3.6 α-cut of Fuzzy Relation

We have learned about α-cut for fuzzy sets, and we know a fuzzy relation is one kind of fuzzy sets. Therefore, we can apply the α-cut to the fuzzy relation.

Definition (α-cut relation) We can obtain α-cut relation from a fuzzy relation by taking the pairs which have membership degrees no less than α. Assume $R \subseteq A \times B$, and R_α is a α-cut relation. Then

$$R_\alpha = \{(x, y) \mid \mu_R(x, y) \geq \alpha, \quad x \in A, y \in B\}$$

Note that R_α is a crisp relation.   □

Example 3.7 For example, we have a fuzzy relation R.

$$M_R = \begin{array}{|ccc}
0.9 & 0.4 & 0.0 \\
0.2 & 1.0 & 0.4 \\
0.0 & 0.7 & 1.0 \\
0.4 & 0.2 & 0.0
\end{array}$$

Now the level set with degrees of membership function is,

$$\Lambda = \{0, 0.2, 0.4, 0.7, 0.9, 1.0\}$$

then we can have some α-cut relations in the following.

$$M_{R\,0.4} = \begin{array}{|ccc}
1 & 1 & 0 \\
0 & 1 & 1 \\
0 & 1 & 1 \\
1 & 0 & 0
\end{array}$$

$$M_{R\,0.7} = \begin{array}{|ccc}
1 & 0 & 0 \\
0 & 1 & 0 \\
0 & 1 & 1 \\
0 & 0 & 0
\end{array}$$

$$M_{R\,0.9} = \begin{array}{|ccc}
1 & 0 & 0 \\
0 & 1 & 0 \\
0 & 0 & 1 \\
0 & 0 & 0
\end{array}$$

$$M_{R\,1.0} = \begin{array}{|ccc}
0 & 0 & 0 \\
0 & 1 & 0 \\
0 & 0 & 1 \\
0 & 0 & 0
\end{array}$$

3.3.7 Projection and Cylindrical Extension

Definition (Decomposition of relation) Fuzzy relation can be said to be composed of several R_α's as following.

$$R = \bigcup_{\alpha} \alpha R_\alpha$$

Here α is a value in the level set; R_α is a α-cut relation; αR_α is a fuzzy relation. The membership function of αR_α is defined as,

$$\mu_{\alpha R_\alpha}(x, y) = \alpha \bullet \mu_{R_\alpha}(x, y) , \quad \text{for } (x, y) \in A \times B$$

Thus we can decompose a fuzzy relation R into several αR_α. □

Example 3.8 The relation R in the previous example can be decomposed as following.

$$M_R = 0.4 \times \begin{vmatrix} 1 & 1 & 0 \\ 0 & 1 & 1 \\ 0 & 1 & 1 \\ 1 & 0 & 0 \end{vmatrix} \cup 0.7 \times \begin{vmatrix} 1 & 0 & 0 \\ 0 & 1 & 0 \\ 0 & 1 & 1 \\ 0 & 0 & 0 \end{vmatrix} \cup 0.9 \times \begin{vmatrix} 1 & 0 & 1 \\ 0 & 1 & 0 \\ 0 & 0 & 1 \\ 0 & 0 & 0 \end{vmatrix} \cup 1.0 \times \begin{vmatrix} 0 & 0 & 0 \\ 0 & 1 & 0 \\ 0 & 0 & 1 \\ 0 & 0 & 0 \end{vmatrix}$$

Definition (Projection) We can project a fuzzy relation $R \subseteq A \times B$ with respect to A or B as in the following manner.
For all $x \in A, y \in B$,

$$\mu_{R_A}(x) = \underset{y}{Max}\, \mu_R(x, y) \quad : \text{projection to A}$$

$$\mu_{R_B}(y) = \underset{x}{Max}\, \mu_R(x, y) \quad : \text{projection to B}$$

Here the projected relation of R to A is denoted by R_A, and to B is by R_B. □

Example 3.9 There is a relation $R \subseteq A \times B$. The projection with respect to A or B shall be,

$$M_R = $$

A \ B	b_1	b_2	b_3
a_1	0.1	0.2	1.0
a_2	0.6	0.8	0.0
a_3	0.0	1.0	0.3

$$M_{R_A} = $$

a_1	1.0
a_2	0.8
a_3	1.0

$$M_{R_B} = $$

b_1	b_2	b_3
0.6	1.0	1.0

In the projection to A, the strongest degree of relation concerning a_1 is 1.0, that for a_2 is 0.8 and that for a_3 is 1.0.   □

Definition (Projection in n dimension) So far has been the projection in 2-dimensions relation. Extending it to n-dimensional fuzzy set, assume relation R is defined in the space of X1 × X2 × ... × Xn. Projecting this relation to subspace of Xi1 × Xi2 × ... × Xik, gives a projected relation :

$$R_{X_{i1} \times X_{i2} \times ... \times X_{ik}}$$

$$\mu_{R_{X_{i1} \times X_{i2} \times ... \times X_{ik}}}(x_{i1}, x_{i2}, ..., x_{ik}) = \underset{X_{j1}, X_{j2}, ..., X_{jm}}{Max}\ \mu_R(x_1, x_2, ..., x_n)$$

Here X_{j1}, X_{j2}, ..., X_{jm} represent the omitted dimensions, and X_{i1}, X_{i2}, ..., X_{ik} the remained dimensions, and thus

$$\{X_1, X_2, ... , X_n\} = \{X_{i1}, X_{i2}, ... , X_{ik}\} \cup \{X_{j1}, X_{j2}, ... , X_{jm}\}$$ □

Definition (Cylindrical extension) As the opposite concept of projection, cylindrical extension is possible. If a fuzzy set or fuzzy relation R is defined in space $A \times B$, this relation can be extended to $A \times B \times C$ and we can obtain a new fuzzy set. This fuzzy set is written as $C(R)$.

$$\mu_{C(R)}(a, b, c) = \mu_R(a, b)$$
$$a \in A, b \in B, c \in C$$ □

Example 3.10 In the previous example, relation R_A is the projection of R to direction A. If we extend it again to direction B, we can have an extended relation $C(R_A)$. For example

$$\mu_{C(R_A)}(a_1, b_1) = \mu_{R_A}(a_1) = 1.0$$
$$\mu_{C(R_A)}(a_1, b_2) = \mu_{R_A}(a_1) = 1.0$$
$$\mu_{C(R_A)}(a_2, b_1) = \mu_{R_A}(a_2) = 0.8$$

$$M_{C(R_A)} = $$

	b_1	b_2	b_3
a_1	1.0	1.0	1.0
a_2	0.8	0.8	0.8
a_3	1.0	1.0	1.0

The new relation $C(R_A)$ is now in $A \times B$.   □

Let two fuzzy relations be defined as follows:

$$R \subseteq X_1 \times X_2, S \subseteq X_2 \times X_3$$

Even though we want to apply the intersection operation between R and S, it is not possible because the domains of R and S are different each other. If we obtain cylindrical extensions $C(R)$ and $C(S)$ to space of $X_1 \times X_2 \times X_3$, and then $C(R)$ and $C(S)$ have the same domain. We can now apply operations on the two extended sets $C(R)$ and $C(S)$. Therefore join (or intersection) of R and S can be calculated by the intersection of $C(R)$ and $C(S)$.

$$\text{join } (R, S) = C(R) \cap C(S)$$

The projection and cylindrical extension are often used to make domains same for more than one fuzzy sets.

3.4 Extension of Fuzzy Set

3.4.1 Extension by Relation

Definition (Extension of fuzzy set) Let A and B be fuzzy sets and R denote the relation from A to B. This relation can be expressed by a function f,

$$x \in A, y \in B$$
$$y = f(x) \quad \text{or} \quad x = f^{-1}(y)$$

Here we used the term "function" without considering the strict condition for being a function. Then we can obtain make fuzzy set B' in B by R and A.

for $y \in B$,

$$\mu_{B'}(y) = \underset{x \in f^{-1}(y)}{Max} [\mu_A(x)] \quad \text{if} \quad f^{-1}(y) \neq \varnothing \quad \square$$

Example 3.11 Let fuzzy set A be "the set of people with an infectious disease" and the crisp set B be "the set of people having been in contact with the infected people". The contact relation is given by R in Fig 3.18

$$A = \{(a_1, 0.4), (a_2, 0.5), (a_3, 0.9), (a_4, 0.6)\}$$
$$B = \{b_1, b_2, b_3\}$$

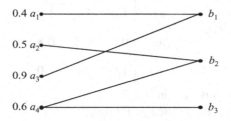

Fig. 3.18. Extension of a fuzzy relation

For example, by A, the possibility of person a_1 is 0.4, and by relation R, a_1 has been in contact with b. With such a set A and relation R, the infectious set B' in B can be obtained as follows :
First for b_1,

$$f^{-1}(b1) = \{(a1, 0.4), (a3, 0.9)\}, \quad Max\ [0.4, 0.9] = 0.9$$
$$\Rightarrow \mu_{B'}(b_1) = 0.9$$

Now for b_1,

$$f^{-1}(b2) = \{(a2, 0.5), (a4, 0.6)\}, \quad Max\ [0.5, 0.6] = 0.6$$
$$\Rightarrow \mu_{B'}(b_2) = 0.6$$

Similarly for b_2,

$$f^{-1}(b3) = \{(a_4, 0.6)\}$$
$$\Rightarrow \mu_{B'}(b_3) = 0.6$$

Arranging all,

$$B' = \{(b_1, 0.9), (b_2, 0.6), (b_3, 0.6)\}, \text{It is the result of inference.} \quad \square$$

3.4.2 Extension Principle

Definition (Extension principle) We can generalize the pre-explained extension of fuzzy set. Let X be Cartesian product of universal set $X = X_1 \times X_2 \times ... \times X_r$ and $A_1, A_2, ... , A_r$ be r fuzzy sets in the universal set.
Cartesian product of fuzzy sets $A_1, A_2, ... , A_r$ yields a fuzzy set $A_1 \times A_2 \times ... \times A_r$ defined as

$$\mu_{A_1 \times A_2 \times ... \times A_r}(x_1 \times x_2 \times ... \times x_r) = \quad Min\ [\ \mu_{A_1}(x_1), ... , \mu_{A_r}(x_r)\ ]$$

Let function f be from space X to Y,
$$f(x_1, x_2, ... , x_r) \quad : \quad X \rightarrow Y$$
Then fuzzy set B in Y can be obtained by function f and fuzzy sets A_1, $A_2, ... , A_r$ as follows:

$$\mu_B(y) = \begin{cases} 0, \text{if } f^{-1}(y) = \varnothing \\ \\ \underset{y = f(x_1, x_2, ..., x_r)}{Max} \left[Min\left(\mu_{A_1}(x_1), ..., \mu_{A_r}(x_r) \right) \right], \text{otherwise} \end{cases}$$

Here, $f^{-1}(y)$ is the inverse image of y. $\mu_B(y)$ is the membership of $y = (x_1, ..., x_r)$ whose membership function is $\mu_{A_1 \times A_2 \times ... \times A_r}(x_1, ..., x_r)$. $\square$
If f is a one-to-one correspondence function, $\mu_B(y) = \mu_A(f^{-1}(y))$, when $f^{-1}(y) \neq \varnothing$.

3.4.3 Extension by Fuzzy Relation

Definition (Extension of fuzzy relation) For given fuzzy set A, crisp set B and fuzzy relation $R \subseteq A \times B$, there might be a mapping function expressing the fuzzy relation R. Membership function of fuzzy set B' in B is defined as follows :

For $x \in A$, $y \in B$, and $B' \subseteq B$

$$\mu_{B'}(y) = \underset{x \in f^{-1}(y)}{\text{Max}}[\text{Min}(\mu_A(x), \mu_R(x, y))] \quad \square$$

Example 3.12 Fig 3.19 shows a further generalization of the example in the previous section. The fuzzy set A stands for the infectious patients, and the set B for the people who have been in contact with those patients. The contact degree is given by the relation R. To determine the fuzzy set B', we need a process of inference. In this inference, note that Max-Min operation is used just as in the composition of two fuzzy relations.

The following shows examples of calculation for the membership of B'.

For b_1,

        Min $[\mu_A(a_1), \mu_R(a_1, b_1)] = \text{Min}[0.4, 0.8] = 0.4$
        Min $[\mu_A(a_3), \mu_R(a_3, b_1)] = \text{Min}[0.9, 0.3] = 0.3$
        Max $[0.4, 0.3] = 0.4$

$$\Rightarrow \mu_{B'}(b_1) = 0.4$$

For b_2,

        Min $[\mu_A(a_2), \mu_R(a_2, b_2)] = \text{Min}[0.5, 0.2] = 0.2$
        Min $[\mu_A(a_4), \mu_R(a_4, b_2)] = \text{Min}[0.6, 0.7] = 0.6$
        Max $[0.2, 0.6] = 0.6$

$$\Rightarrow \mu_{B'}(b_2) = 0.6$$

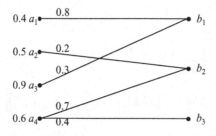

Fig. 3.19. Extension by fuzzy relation

For b_3,

$$\text{Max Min } [\mu_A (a_4), \mu_R (a_4, b_3)] = \text{Max Min } [0.6, 0.4] = 0.4$$
$$\Rightarrow \mu_{B'} (b_3) = 0.4$$

So fuzzy set B' obtained by fuzzy set A and fuzzy relation R is,
$B' = \{(b_1, 0.4), (b_2, 0.6), (b_3, 0.4)\}$ □

Extension of fuzzy set and fuzzy relation is also possible among the several relations and sets. That is, the fuzziness in fuzzy set A can be propagated through more than one relations and sets.

Example 3.13 Fuzzy set A can make a fuzzy set B' in crisp set B by fuzzy relation $R_1 \subseteq A \times B$, and B' again can make a fuzzy set C' from fuzzy relation $R_2 \subseteq B \times C$.

$$A = \{(a_1, 0.8), (a_2, 0.3)\}$$
$$B = \{b_1, b_2, b_3\}$$
$$C = \{c_1, c_2, c_3\}$$

$M_{R_1} =$

	b_1	b_2	b_3
a_1	0.3	1.0	0.0
a_2	0.8	0.0	0.0

$M_2 =$

	c_1	c_2	c_3
b_1	0.7	0.4	1.0
b_2	0.2	0.0	0.8
b_3	0.0	0.3	0.9

By fuzzy set A and fuzzy relation R_1, we get the following B'.
$$B' = \{(b_1, 0.3), (b_2, 0.8), (b_3, 0)\}$$
Again by B' and R_2, we get C'
$$C' = \{(c_1, 0.3), (c_2, 0.3), (c_3, 0.8)\}$$
The original fuzziness in the set A was propagated to C' through B, C, R_1 and R_2 □

3.4.4 Fuzzy Distance between Fuzzy Sets

Let the universal set X be a certain space where pseudo-metric distance d is applied, and then following characteristics are satisfied.

Definition (pseudo-metric distance) If d is a mapping function from space X^2 to $\mathcal{R}^+$ (positive real numbers) and holds the following properties, it is called pseudo-metric distance
   i) $d(x, x) = 0, \forall x \in X$
   ii) $d(x_1, x_2) = d(x_2, x_1), \forall x_1, x_2 \in X$
   iii) $d(x_1, x_3) \leq d(x_1, x_2) + d(x_2, x_3), \forall x_1, x_2, x_3 \in X$ □

If the following condition (4) is added to the pseudo-metric distance, it becomes a metric distance.

iv) if $d(x_1, x_2) = 0$, then $x_1 = x_2$.

Definition (Distance between fuzzy sets) In space X, pseudo-metric distance $d(A, B)$ between fuzzy sets A and B can be defined by extension principle. The distance $d(A, B)$ is given as a fuzzy set.

$$\forall\ \delta \in \mathfrak{R}^{+},\ \mu_{d(A,\ B)}(\delta) = \underset{\delta = d(a,\ b)}{\text{Max}}[\text{Min}\ (\mu_A(a),\ \mu_B(b))] \qquad \square$$

Example 3.14 (Fig 3.20) shows the fuzzy distance between the fuzzy sets $A = \{(1, 0.5), (2, 1), (3, 0.3)\}$ and $B = \{(2, 0.4), (3, 0.4), (4, 1)\}$. (Table 3.2) shows the calculation procedure for $d(A,B)$ from A and B.

Table 3.2. Calculation of fuzzy set d(A,B)

$\delta \in$ d(A, B)	$a \in A$	$b \in B$	$\mu_A(a)$	$\mu_B(b)$	Min	Max, $\mu(\delta)$, d(A, B)
0	2	2	1.0	0.4	0.4	0.4
	3	3	0.3	0.4	0.3	
1	1	2	0.5	0.4	0.4	0.4
	2	3	1.0	0.4	0.4	
	3	2	0.3	0.4	0.3	
	3	4	0.3	1.0	0.3	
2	1	3	0.5	0.4	0.4	1.0
	2	4	1.0	1.0	1.0	
3	1	4	0.5	1.0	0.5	0.5

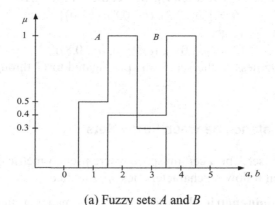

(a) Fuzzy sets A and B

Fig. 3.20. Fuzzy distance set

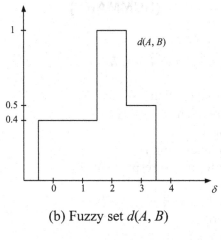

(b) Fuzzy set $d(A, B)$

Fig. 3.20. (cont')

[SUMMARY]

☐ Cartesian product
$$A \times B = \{(a, b) \mid a \in A, b \in B\}$$

☐ Characteristics of a relation
  – One-to-many relation
  – Surjection (onto)
  – Injection (into, one-to-one)
  – Bijection (one-to-one correspondence)

☐ Operations on relations $R, S \subseteq A \times B$
  – Union of relations
  – Intersection of relations
  – Complement of a relation
  – Inverse of a relation
  – Composition of relations

☐ Path and connectivity relation
  – Path of length n
  – Reachability
  – Connectivity

☐ Properties of relation $R \subseteq A \times A$
  – Reflexive relation, irreflexive relation, antireflexive
  – Symmetric relation, antisysmmetric relation, asymmetric relation
  – Transitive relation

☐ Equivalence relation
  – Reflexive relation
  – Symmetric relation
  – Binary relation

☐ Compatible relation
  – Reflexive relation
  – Symmetric relation

☐ Pre-order relation
  – Reflexive relation

- Symmetric relation

☐ Order relation
- Reflexive relation
- Non Symmetric relation
- Binary relation

☐ Fuzzy relation
- Fuzzy relation is a kind of fuzzy set
- Fuzzy relation is given by a fuzzy restriction to a crisp relation

☐ Fuzzy matrix
- One representation of fuzzy relation
- Operations of sum, max product and scalar product

☐ Operation of fuzzy relation
- Fuzzy union relation
- Fuzzy intersection relation
- Fuzzy complement relation
- Fuzzy inverse
- Composition of fuzzy relation

☐ Composition of fuzzy relation
- $S \bullet R$ where $R \subseteq A \times B,\ S \subseteq B \times C$
- Max-min operation is popular
- One kind of inference

☐ α-cut relation
- Because relation is a fuzzy set
- It gives a crisp relation

☐ Projection and cylindrical extension
- Reducing and increasing the number of dimension
- Useful for coinciding the domains of two sets

☐ Extension principle
- Propagation of fuzziness by fuzzy relation
- Extension by fuzzy set and fuzzy relation

☐ Fuzzy distance
- Distance between fuzzy sets represented by a fuzzy set
- One kind of extension of fuzziness

[EXERCISES]

3.1 Find the transitive closure for $A = \{a, b, c, d\}$ and $R = \{(a, b), (b, c), (c, d), (d, b)\}$.

3.2 Obtain a partition of the set $A = \{a, b, c, d, e\}$ by the equivalence relation R.

R	a	b	c	d	e
a	1	1			1
b	1	1			1
c			1	1	
d			1	1	
e	1	1			1

3.3 Compute the complements, intersection and union of the following fuzzy relations R and S.

R	a	b	c	d
a	1.0	0.2	0.4	0.0
b	0.0	0.1	0.0	0.9
c	0.1	0.0	1.0	0.0
d	0.0	0.4	0.0	1.1

S	a	b	c	d
a	1.0	0.0	0.0	0.0
b	0.0	0.0	0.4	0.9
c	0.4	0.0	0.1	0.0
d	0.5	1.0	0.0	0.0

3.4 Determine the composition relation $S \bullet R \subseteq A \times C$ where $R \subseteq A \times B$ and $S \subseteq B \times C$ are defined as follows

R	a	b	c	d
1	0.4	0.0	0.0	1.0
2	0.5	0.4	0.9	0.0
3	0.2	0.1	1.0	0.4
4	0.0	0.2	0.0	1.0

S	a	b	c
a	0.4	0.1	0.0
b	0.2	0.0	0.9
c	0.2	0.0	0.5
d	0.1	0.0	0.9

3.5 Determine the α-cut relation for the following fuzzy relation where $\alpha =$ 0.4 and 0.8.

R	1	2	3	4
a	0.4	0.0	0.5	0.8
b	0.4	0.0	0.9	0.1
c	0.0	0.4	0.0	0.2
d	0.0	0.8	0.0	0.0

3.6 Consider a fuzzy set A and a crisp set B
$$A = \{(x, 0.4), (y, 0.9), (z, 1.0), (w, 0.1)\}$$
$$B = \{a, b, c\}$$
Determine a fuzzy set $B' \subseteq B$ induced by A and the relation $R \subseteq A \times B$.

R	a	b	c
x	0.0	0.4	0.8
y	0.9	0.9	0.7
z	1.0	0.0	0.5
w	0.0	0.1	0.8

3.7 Consider a fuzzy relation $R \subseteq A \times A$ where $A = \{a, b, c\}$.

R	a	b	c	d	e
a	1	1	0	0	1
b	0	1	1	0	0
c	0	0	1	0	0
d	0	0	0	1	0
e	0	0	1	1	1

a) Determine the characteristic of this relation.
b) Show an ordinal function of the relation if it is an order relation.

3.8 Determine the fuzzy set B induced by A and $f(x) = x^2$

$$A = \{(-2, 0.8), (-5, 0.5), (0, 0.8), (1, 1.0), (2, 0.4), (3, 0.1)\}$$

$$B = \left\{ y \mid y = f(x), \mu_B(y) = \underset{x:y=f(x)}{Max} \ \mu_A(x) \right\}$$

3.9 There are fuzzy set A, crisp set B and fuzzy relation R

$A = \{(x, \mu_A(x)) \mid 0 \leq x \leq 1, \mu_A(x) = x^2\}$
$B = \{(y, \mu_B(y)) \mid 0 \leq y \leq 1\}$
$R = \{((x, y), \mu_R(x, y)) \mid 0 \leq x + y = 1, x \in A, y \in B\}$
where $\mu_R(x, y) = \min[x^2, y^2]$

Determine the fuzzy set $B' \subseteq B$ induced by A, B and R

3.10 There is a relation R_{123}. $R_{123} \subset X_1 \times X_2 \times X_3 = 0.9 / (x, a, \alpha) + 0.4 / (x, b, \alpha) + 1.0 / (y, a, \alpha) + 0.7 / (y, a, \beta)$

$R_{123} \subset X_1 \times X_2 \times X_3$ where $X_1 = \{x, y\}$, $X_2 = \{a, b\}$, $X_3 = \{\alpha, \beta\}$

a) Determine $R_{12} \subset X_1 \times X_2$ and $R_{23} \subset X_2 \times X_3$ by projection.
b) Obtain R_{123} by cylindrical extension of R_{12} and R_{23}.
c) Obtain $R_{1234} \subseteq X_1 \times X_2 \times X_3 \times X_4$ by cylindrical extension where $X_4 = \{p,q\}$

Chapter 4. FUZZY GRAPH AND RELATION

Based on the concepts of fuzzy relation described in the previous chapter, we introduce fuzzy graph and it's related topics. We will also develop characteristics of fuzzy relation and study various types of fuzzy relations. The concepts of fuzzy homomorphism and strong homomorphism are also introduced.

4.1 Fuzzy Graph

4.1.1 Graph and Fuzzy Graph

Definition (Graph) A graph G is defined as follows.

$G = (V, E)$
V : Set of **vertices**. A vertex is also called a node or element.
E : Set of **edges**. An edge is pair (x, y) of vertices in V. □

A graph is a data structure expressing relation $R \subseteq V \times V$. When order in pair (x, y) is defined, the pair is called edge with direction and we call such graph directed graph. When order is not allowed, we call it undirected graph.

Path from x to y is a set of edges when continuous edges of (x, a_1), (a_1, a_2), (a_2, a_3), ... , (a_n, y) are existing. Length of path is a the number of edges in this path. When there exists a path from node a to b in G, a and b are said to be connected. If all $a, b \in V$ in graph G are connected, this graph is said to be a connected graph.

When sets A and B are given (including the case $A = B$), let's define a crisp relation $R \subseteq A \times B$. For $x \in A$, $y \in B$, if $(x, y) \in R$, there exists an edge between x and y. In other words,

$$\forall (x, y) \in R \Leftrightarrow \mu_R(x, y) = \mu_G(x, y) = 1$$

Here given the relation R is a fuzzy relation, and the membership function $\mu_R(x, y)$ enables $\mu_G(x, y)$ value to be between 0 and 1. Such graph is called a fuzzy graph.

Definition (Fuzzy graph)

$$\tilde{G} = \left(V, \, \tilde{E} \right)$$

V : set of vertices
E : fuzzy set of edges between vertices   □

We can think of set V which is a fuzzy set. In this case, we say this graph represents fuzzy relation of fuzzy nodes , and can be defined as following.

$$\tilde{G} = \left(\tilde{V}, \, \tilde{E} \right)$$

In this book, we replace $\tilde{G} = \left(V, \, \tilde{E} \right)$ with $G = (V, E)$ for convenience.

4.1.2 Fuzzy Graph and Fuzzy Relation

Fuzzy graph is an expression of fuzzy relation and thus the fuzzy graph is frequently expressed as fuzzy matrix.

Example 4.1 Fig 4.1 shows an example of fuzzy graph represented as fuzzy relation matrix M_G.   □

M_G	b_1	b_2
a_1	0.8	0.2
a_2	0.3	0.0
a_3	0.7	0.4

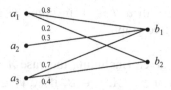

Fig. 4.1. Fuzzy graph

Example 4.2 For instance, let $\mathscr{R}^+$ be nonnegative real numbers. For $x \in \mathscr{R}^+$ and $y \in \mathscr{R}^+$, a relation R is defined as,

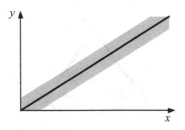

Fig. 4.2. Fuzzy graph $y \approx x$

$$R = \{(x, y) \mid x \approx y\}, R \subseteq \mathscr{R}^+ \times \mathscr{R}^+.$$

The symbol $\approx$ represents " close to "(Fig 4.2).   □

Example 4.3 Assume that there is a set $A = \{a_1, a_2, a_3\}$, and fuzzy relation R is defined in $A \times A$. The fuzzy relation R is illustrated in Fig 4.3(a). In the figure, the darkness of color stands for the strength of relation. Relation (a, b) is stronger than that of relation (a, c). The corresp-onding fuzzy graph is shown in Fig 4.3(b). Here the strength of relation is marked by the thickness of line.   □

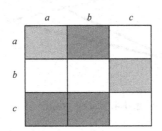

(a) Fuzzy relation R

Fig. 4.3. Fuzzy relation and fuzzy graph

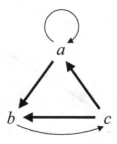

(b) Fuzzy graph

Fig. 4.3. (cont')

Example 4.4 When a fuzzy relation is defined by the following fuzzy matrix, this corresponding fuzzy graph can be depicted in Fig 4.4. Note the thickness of edge represents strength of relation.

	b_1	b_2	b_3
a_1	0.5	1.0	0.0
a_2	0.0	0.0	0.5
a_3	1.0	1.0	0.0

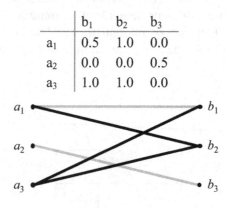

Fig. 4.4. Fuzzy graph

If we make mapping function $\Gamma(A)$ represent the corresponding set of A, we get the following fuzzy sets from this fuzzy relation. That is, we obtain new fuzzy set in the set B by the set A and fuzzy relation R.

$\Gamma\{a_1\} = \{(b_1, 0.5), (b_2, 1.0)\}$
$\Gamma\{a_2\} = \{(b_3, 0.5)\}$
$\Gamma\{a_3\} = \{(b_1, 1.0), (b_2, 1.0)\}$
$\Gamma\{a_1, a_2\} = \{(b_1, 0.5), (b_2, 1.0), (b_3, 0.5)\}$ □

Example 4.5 Fig 4.5 might be thought to be a picture received by a camera. This picture is drawn on two dimensional coordinates. If we associate

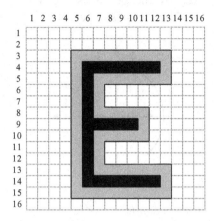

Fig. 4.5. Fuzzy graph (picture)

fuzzy relation $\mu_R(x, y)$ with each coordinate's darkness, this figure can be equivalent to a fuzzy graph. Also Fig 4.6 is for two-dimensional coordinates and we can also associate membership degree $\mu_R(x, y)$ to the darkness of each point(x, y). □

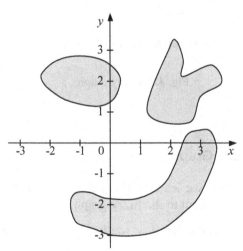

Fig. 4.6. Fuzzy graph (by coordinates)

Example 4.6 When $\mathscr{R}$ is the real number, let $R \subseteq \mathscr{R} \times \mathscr{R}$ is defined as follows, for $x \in \mathscr{R}, y \in \mathscr{R}$,
$$\mu_R(x, y) = x^2 + y^2 = 1$$
This relation is expressed by the ordinary figure in (Fig 4.7(a)). And if
$$\mu_R(x, y) = x^2 + y^2 \le 1$$

the relation is represented by the graph as Fig 4.7(b). The intensity of
relation (x, y) is represented by darkness of color.   □

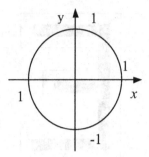

(a) Graph $\mu_R(x, y) = x^2 + y^2 = 1$

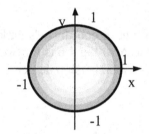

(b) Graph $\mu_R (x, y) = x^2 + y^2 \leq 1$

Fig. 4.7. Fuzzy graph

4.1.3 α-cut of Fuzzy Graph

In the previous chapter, we learned about the a-cut fuzzy relation . It is
natural to extend this concept to the fuzzy graph.

Example 4.7 For example, for $A = \{a, b, c\}$, $R \subseteq A \times A$ is defined as
follows.

$$
M_R = \quad
\begin{array}{c|ccc}
 & a & b & c \\
\hline
a & 1.0 & 0.8 & 0.4 \\
b & 0.0 & 0.4 & 0.0 \\
c & 0.8 & 1.0 & 0.0 \\
\end{array}
$$

For level set $\{0, 0.4, 0.8, 1\}$, if we apply the α-cut operation, we also get crisp relations and corresponding graphs in Fig 4.8.   □

If we denote the graph from α-cut as G_α, the following relation between α and G_α holds.

$$\alpha_1 \geq \alpha_2 \;\Rightarrow\; \begin{array}{l} R_{\alpha_1} \subseteq R_{\alpha_2} \\ G_{\alpha_1} \subseteq G_{\alpha_2} \end{array}$$

$$M_R = \begin{array}{c|ccc} & a & b & c \\ \hline a & 1.0 & 0.8 & 0.4 \\ b & 0.0 & 0.4 & 0.0 \\ c & 0.8 & 1.0 & 0.0 \end{array}$$

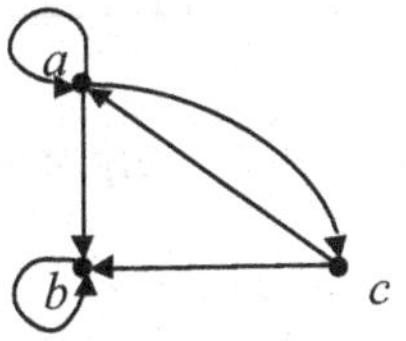

$$M_{R_{0.4}} = \begin{array}{c|ccc} & a & b & c \\ \hline a & 1.0 & 1.0 & 1.0 \\ b & 0.0 & 1.0 & 0.0 \\ c & 1.0 & 1.0 & 0.0 \end{array}$$

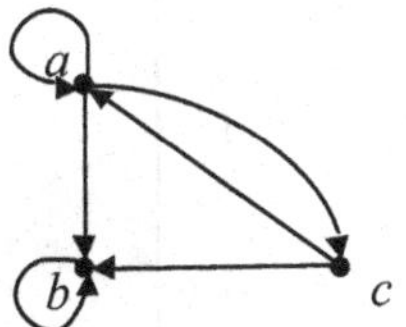

$$M_{R_{0.8}} = \begin{array}{c|ccc} & a & b & c \\ \hline a & 1.0 & 1.0 & 0.0 \\ b & 0.0 & 0.0 & 0.0 \\ c & 1.0 & 1.0 & 0.0 \end{array}$$

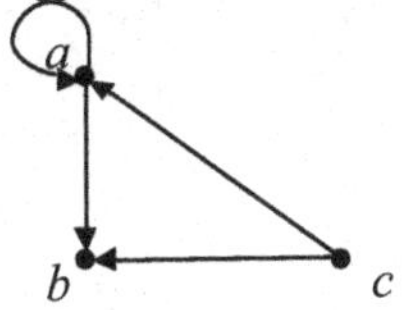

$$M_{R_{1.0}} = \begin{array}{c|ccc} & a & b & c \\ \hline a & 1.0 & 0.0 & 0.0 \\ b & 0.0 & 0.0 & 0.0 \\ c & 0.0 & 1.0 & 0.0 \end{array}$$

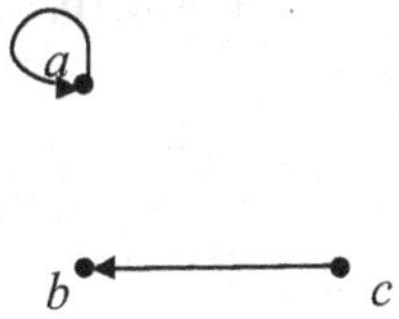

Fig. 4.8. α-cut of fuzzy graph

Example 4.8 There is a relation defined as follows:

$$\mu_R(x, y) = x/2 + y \leq 1.$$

Show graphical representations of R and its α-cut relation $R_{0.5}$ at $\alpha=0.5$ Figs 4.9 and 4.10 show R and $R_{0.5}$. The density of darkness represents the membership degrees

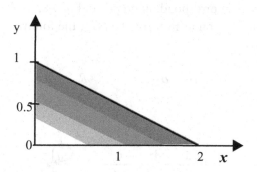

Fig. 4.9. Graphical form of R

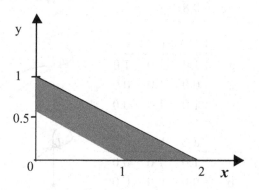

Fig. 4.10. Graphical representation of $R_{0.5}$

Example 4.9 Show the following set A and relation R.

i) $\mu_A(x) = x$

ii) $\mu_R(x,y) = x+y \le 1, \quad x \in A, \quad 0 \le y \le 1.$

The set A and R are shown in Figs 4.11, 4.12, respectively.

Example 4.10 There is a set $A \subset \Re$ where $\Re$ is the set of real numbers and $x \in \Re$. A={ x | x close to $2k\pi$, k = -1,0, 1,2,....}. The membership function of the set A is formally defined as $\mu_A(x) = \text{Max}[0, cosx]$.

i)   Show the graphical representation of A (Fig 4.13).

ii)   Show the α-cut set of A at α=0.5 (Fig 4.14).

iii) Show the relation defined as follows:

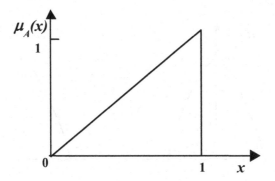

Fig. 4.11. Set $\mu_A(x) = x$

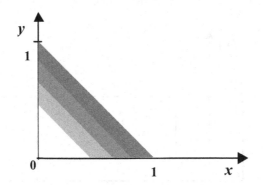

Fig. 4.12. Relation $\mu_A(x,y) = x+y \leq 1$, $x \in A$.

iv)   Show the relation defined as follows:
$R = \{(x,y) \mid y = cos\ x \geq 0,\quad x \in A\}$ (Fig 4.15).

v)   Show the α-cut relation of R at $\alpha = 0.5$ (Fig 4.16).

vi)   Show the set defined by $\mu_B(y) = cosx$, $x \in A$. This set B is a set induced by R and A. (Fig 4.17).

vii)   Show the relation defined as follows: $\mu_R(x,y) = Max[0,\ sinx]$, $x \in A$ (Fig 4.18).

The solutions are given in Figs 4.13 – 4.18. In the figures, the membership degrees are represented by the density of darkness.

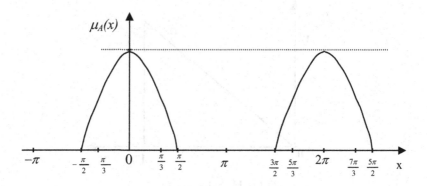

Fig. 4.13. Set $\mu_A(x)=cosx \geq 0$

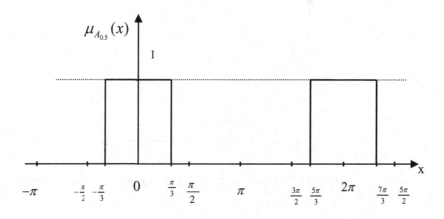

Fig. 4.14. α-cut set $A_{0.5}$

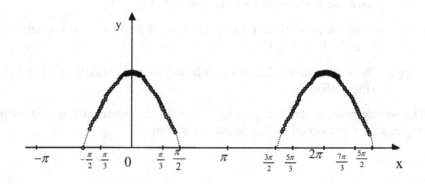

Fig. 4.15. Relation $\mu_R(x,y)=cosx$.

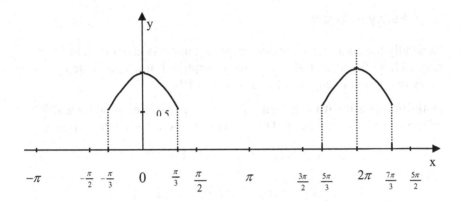

Fig. 4.16. α-cut relation R$_{0.5}$

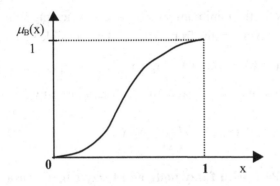

Fig. 4.17. $\mu_B(y)=cos\ x,\ x\in$ A.

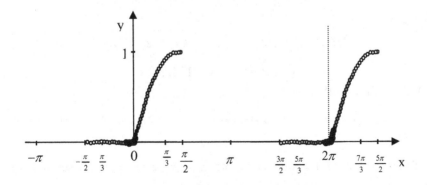

Fig. 4.18. $\mu_R\ (x,y)=$max[0, $sinx$]

4.1.4 Fuzzy Network

Generally the connected directed graph is named as network. This network may well be fuzzified just as the fuzzy graph did. It can be done by giving fuzzy restriction to edges and nodes (Fig. 4.19).

Definition (Path with fuzzy edge) Let V be a crisp set of nodes and R be a relation defined on the set V. Denote the path from node x_{i_1} to x_{ir} by C_i.

$$C_i = (x_{i_1}, x_{i_2}, \ldots , x_{i_r}), x_{i_k} \in V, k = 1, 2, \ldots , r$$

where

$$\forall (x_{i_k}, x_{i_{k+1}}),\ \mu_R(x_{i_k}, x_{i_{k+1}}) > 0, k = 1, 2, \ldots , r\text{-}1$$

We define fuzzy value l for path C_i with

$$l(x_{i_1}, x_{i_2}, \ldots , x_{i_r}) = \mu_R(x_{i_1}, x_{i_2}) \wedge \mu_R(x_{i_2}, x_{i_3}) \wedge \ldots \wedge \mu_R(x_{i_{r-1}}, x_{i_r})$$

This value is for the minimum possibility of connecting from x_{i_1} to x_{i_r}.

If there lie several paths from x_{i_1} to x_{i_r}, then possible set of paths is

$$C(x_i, x_j) = \{c(x_i, x_j) \mid c(x_i, x_j) = (x_{i_1} = x_i, x_{i_2}, \ldots , x_{i_r} = x_j)\}$$

among the possible paths, value of maximum intensity path l^* is gathered as,

$$l^* (x_i, x_j) = \underset{C(x_i, x_j)}{\vee} l \ (x_{i_1} = x_i, x_{i_2}, \ldots , x_{i_r} = x_j) \qquad \square$$

Definition (Path with fuzzy node and fuzzy edge) Consider fuzzy set V of nodes and that of edge R. Then the path C_i from x_{i_1} to x_{i_r} is,

$$C_i = (x_{i_1}, x_{i_2}, \ldots , x_{i_r}), x_{i_k} \in V, k = 1, 2, \ldots , r$$

where,

$$\forall (x_{i_k}, x_{i_{k+1}}),\ \mu_R(x_{i_k}, x_{i_{k+1}}) > 0, k = 1, 2, \ldots , r\text{-}1$$

$$\forall x_{i_k},\ \mu_V(x_{i_k}) > 0, k = 1, 2, \ldots , r$$

Value of this path is

$$l(x_{i_1}, x_{i_2}, \ldots , x_{i_r}) = \mu_R(x_{i_1}, x_{i_2}) \wedge \mu_R(x_{i_2}, x_{i_3}) \wedge \ldots \wedge \mu_R(x_{i_{r-1}}, x_{i_r}) \wedge \mu_V(x_{i_1}) \wedge \mu_V(x_{i_2}) \wedge \ldots \wedge \mu_V(x_{i_r})$$

If there are more than one paths form x_{i_1} to x_{i_r}, we can get the set of paths and the maximum intensity path on the same way. $\square$

4.2 Characteristics of Fuzzy Relation

In the previous chapter, we have seen reflexive, symmetric and transitive relations. Our main goal is to develop such features for the fuzzy relation. We assume that fuzzy relation R is defined on $A \times A$.

4.2.1 Reflexive Relation

For all $x \in A$, if $\mu_R(x, x) = 1$, we call this relation reflexive.

Example 4.8 For instance, for set $A = \{2, 3, 4, 5\}$ there is a relation such that :

For $x, y \in A$, "x is close to y"

Concerning this relation, when $x = y$, the relation is perfectly satisfied and thus $\mu_R(x, x) = 1$. Let's denote this reflexive one in a matrix as in the following.   □

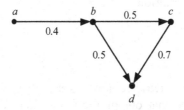

(a) Fuzzy network (edge)

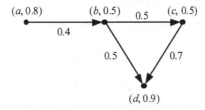

(b) Fuzzy network (node, edge)

Fig. 4.19. Fuzzy network

If $\exists\, x \in A$, $\mu_R(x, y) \neq 1$, then the relation is called " irreflexive " and if $\forall\, x \in A$, $\mu_R(x, y) \neq 1$, then it is called " antireflexive ".

A \ A	2	3	4	5
2	1.0	0.9	0.8	0.7
3	0.9	1.0	0.9	0.8
4	0.8	0.9	1.0	0.9
5	0.7	0.8	0.9	1.0

4.2.2 Symmetric Relation

When fuzzy relation R is defined on $A \times A$, it is called symmetric if it satisfies the following condition.

$$\forall (x, y) \in A \times A$$

$$\mu_R(x, y) = \mu \quad \Rightarrow \quad \mu_R(y, x) = \mu$$

If we express this symmetric relation as a matrix, we get a symmetric matrix. So we easily see that our previous relation "x is close to y" is a symmetric relation.

We say " antisymmetric " for the following case.

$$\forall (x, y) \in A \times A, x \neq y$$

$$\mu_R(x, y) \neq \mu_R(y, x) \text{ or } \mu_R(x, y) = \mu_R(y, x) = 0$$

We can also define the concept of " asymmetric " or " nonsymmetic " as follows.

$$\exists (x, y) \in A \times A, x \neq y$$

$$\mu_R(x, y) \neq \mu_R(y, x)$$

"Perfect antisymmetry" can be thought to be the special case of antisymmetry satisfying :

$$\forall (x, y) \in A \times A, x \neq y$$

$$\mu_R(x, y) > 0 \Rightarrow \mu_R(y, x) = 0$$

Example 4.9 This is an example of perfect antisymmetric relation.

A \ A	a	b	c	d
a	0.0	0.8	0.4	0.6
b	0.0	0.3	0.0	0.6
c	0.0	0.3	1.0	0.2
d	0.0	0.0	0.0	0.8

4.2.3 Transitive Relation

Transitive relation is defined as,

$$\forall (x, y), (y, x), (x, z) \in A \times A$$

$$\mu_R(x, z) \geq \underset{y}{\text{Max}}[\text{Min}(\mu_R(x, y), \mu_R(y, z))]$$

If we use the symbol $\vee$ for Max and $\wedge$ for Min, the last condition becomes.

$$\mu_R(x, z) \geq \underset{y}{\vee}[\mu_R(x, y) \wedge \mu_R(y, z)]$$

If the fuzzy relation R is represented by fuzzy matrix M_R, we know that left side in the above formula corresponds to M_R and right one to M_{R^2}. That is, the right side is identical to the composition of relation R itself. So the previous condition becomes,

$$M_R \geq M_{R^2} \text{ or } R \supseteq R^2$$

Example 4.10 Let's have a look at the relation by fuzzy matrix M_R and it's M_{R^2} obtained by Max-Min composition operation(Fig 4.20).

$M_R =$

	a	b	c
a	0.2	1.0	0.4
b	0.0	0.6	0.3
c	0.0	1.0	0.3

$M_{R^2} =$

	a	b	c
a	0.2	0.6	0.3
b	0.0	0.6	0.3
c	0.0	0.6	0.3

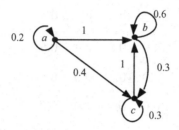

Fig. 4.20. Fuzzy relation (transitive relation)

To investigate whether this is a transitive relation or not, we have to check all elements in M_R and M_{R^2}.

For example,

for (a, a), we have $\mu_R(a, a) \geq \mu_{R^2}(a, a)$

for (a, b), $\mu_R(a, b) \geq \mu_{R^2}(a, b)$

we see $M_R \geq M_{R^2}$ or $R \supset R^2$. Therefore, this fuzzy relation R has the transitive characteristic.

4.2.4 Transitive Closure

As we have referred the expression of fuzzy relation by matrix M_R, fuzzy matrix M_{R^2} corresponding composition R^2 shall be calculated by the multi-plication of M_R.

$$\mu_{R^2}(x, z) = M_R \bullet M_R = \underset{y}{\text{Max}}[\text{Min}(\mu_R(x, y), \mu_R(y, z))]$$

Transitive relation was referred to as $R \supseteq R^2$ and thus the relation between M_R and M_{R^2} holds

$$M_R \geq M_{R^2}$$

then again, the relation $R \supseteq R^3$ may well be satisfied, and by the method of generalization we know

$$R \supseteq R^k \quad k = 1, 2, 3 \dots$$

from the property of closure, the transitive closure of R shall be,

$$R^{\infty} = R \cup R^2 \cup R^3 \cup \dots$$

Generally, if we go on multiplying fuzzy matrices(i.e, composition of r-elation), the following equation is held.

$$R^k = R^{k+1}, \quad k \leq n$$

where $R \subseteq A \times A$ and the cardinality of A is n.
so, R^∞ is easily obtained

$$R^\infty = R \cup R^2 \cup R^3 \cup \ldots \cup R^k, \quad k \le n$$

Example 4.11 There are R, R^2, R^3 represented by matrices in the following.

$$M_R = \begin{array}{c|ccc} & a & b & c \\ \hline a & 0.2 & 1.0 & 0.4 \\ b & 0.0 & 0.6 & 0.3 \\ c & 0.0 & 1.0 & 0.3 \end{array}$$

$$M_{R^2} = \begin{array}{c|ccc} & a & b & c \\ \hline a & 0.2 & 0.6 & 0.3 \\ b & 0.0 & 0.6 & 0.3 \\ c & 0.0 & 0.6 & 0.3 \end{array}$$

$$M_{R^3} = \begin{array}{c|ccc} & a & b & c \\ \hline a & 0.2 & 0.6 & 0.3 \\ b & 0.0 & 0.6 & 0.3 \\ c & 0.0 & 0.6 & 0.3 \end{array}$$

Here $M_R \ge M_{R^2}$ so $R \supseteq R^2$ and transitive relation is verified. Since $M_{R^2} = M_{R^3}$, we know $M_{R^2} = M_{R^3} = \cdots$. Therefore, the transitive closure shall be, $R^\infty = R \cup R^2$ or $M_{R^\infty} = M_R + M_{R^2}$. Note that the sum operation "+" implies Max operator introduced in (sec 3.3.3).

$$M_{R^\infty} = \begin{array}{c|ccc} & a & b & c \\ \hline a & 0.2 & 1.0 & 0.4 \\ b & 0.0 & 0.6 & 0.3 \\ c & 0.0 & 1.0 & 0.3 \end{array} \; + \; \begin{array}{c|ccc} & a & b & c \\ \hline a & 0.2 & 0.6 & 0.3 \\ b & 0.0 & 0.6 & 0.3 \\ c & 0.0 & 0.6 & 0.3 \end{array}$$

$$= \begin{array}{c|ccc} & a & b & c \\ \hline a & 0.2 & 1.0 & 0.4 \\ b & 0.0 & 0.6 & 0.3 \\ c & 0.0 & 1.0 & 0.3 \end{array}$$

4.3 Classification of Fuzzy Relation

We generalize here the concepts of equivalence, compatibility, pre-order and order relations of crisp relations to those of fuzzy relations. We assume relation R is defined on $A \times A$.

4.3.1 Fuzzy Equivalence Relation

Definition (Fuzzy equivalence relation) If a fuzzy relation $R \subseteq A \times A$ satisfies the following conditions, we call it a "fuzzy equivalence relation" or "similarity relation."
   i)   Reflexive relation
$$\forall x \in A \Rightarrow \mu_R(x, x) = 1$$
   ii) Symmetric relation
$$\forall (x, y) \in A \times A, \ \mu_R(x, y) = \mu \Rightarrow \mu_R(y, x) = \mu$$
   iii) Transitive relation
$$\forall (x, y), (y, z), (x, z) \in A \times A$$
$$\mu_R(x, z) \ \geq \ \underset{y}{\mathrm{Max}}[\mathrm{Min}[\mu_R(x, y), \mu_R(y, z)]] \quad \square$$

Example 4.12 Let's consider a fuzzy relation expressed in the following matrix. Since this relation is reflexive, symmetric and transitive, we see that it is a fuzzy equivalence relation (Fig 4.21). $\square$

	a	b	c	d
a	1.0	0.8	0.7	1.0
b	0.8	1.0	0.7	0.8
c	0.7	0.7	1.0	0.7
d	1.0	0.8	0.7	1.0

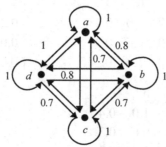

Fig. 4.21. Graph of fuzzy equivalence relation.

Using this similarity relation, we can perform the following three applications.

(1) Partition of sets
Just like crisp set A is done partition into subsets A_1, A_2, ... by the equivalence relation, fuzzy set A also can be performed partition.

Example 4.13 (Fig 4.22) shows a partition of A by the given relation R. At this point, fuzzy equivalence relation holds in class A_1 and A_2, but not between A_1 and A_2. □

	a	b	c	d	e
a	1.0	0.5	1.0	0.0	0.0
b	0.5	1.0	0.5	0.0	0.0
c	1.0	0.5	1.0	0.0	0.0
d	0.0	0.0	0.0	1.0	0.5
e	0.0	0.0	0.0	0.5	1.0

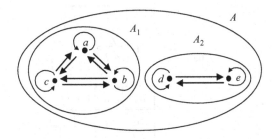

Fig. 4.22. Partition by fuzzy equivalence relation

(2) Partition by α-cut
If α-cut is done on a fuzzy relation, we get crisp relations. By performing α-cut on fuzzy equivalence relation, we get crisp equivalence relations and thus the set A can be partitioned. For instance, if a partition is done on set A into subsets A_1, A_2, A_3, ... , the similarity among elements in A_i is no less than α. The α-cut equivalence relation R_α is defined by

$$\mu_R(x, y) = 1 \text{ if } \mu_R(x, y) \geq \alpha, \forall x, y \in A_i$$
$$= 0 \text{ otherwise}$$

If we apply α-cut according to α_1 in level set $\{\alpha_1, \alpha_2, ... \}$, the partition by this procedure is denoted by $\pi(R_{\alpha_1})$ or $\pi(A/R_{\alpha_1})$. In the same manner, we get $\pi(R_{\alpha_2})$ by the procedure of α_2-cut. Then, we know if $\alpha1 \geq \alpha2$, $R\alpha1 \subseteq R\alpha2$ and we can say that $\pi(R_{\alpha_1})$ is more refined than $\pi(R_{\alpha_2})$.

Example 4.13 Partition tree shows the multiple partitions by α (Fig 4.23).

For instance, if we apply $\alpha = 0.5$, similarity classes are obtained as $\{a, b\}$, $\{d\}$, $\{c, e, f\}$ whose elements have degrees no less than 0.5.

$$\pi(A/R_{0.5}) = \{\{a, b\}, \{d\}, \{c, e, f\}\} \quad \square$$

	a	b	c	d	e	f
a	1.0	0.8	0.0	0.4	0.0	0.0
b	0.8	1.0	0.0	0.4	0.0	0.0
c	0.0	0.0	1.0	0.0	1.0	0.5
d	0.4	0.4	0.0	1.0	0.0	0.0
e	0.0	0.0	1.0	0.0	1.0	0.5
f	0.0	0.0	0.5	0.0	0.5	1.0

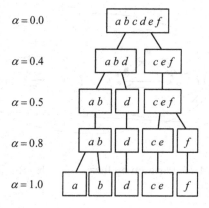

Fig. 4.23. Partition tree

(3) Set similar to element x

If similarity relation R is defined on set A, elements related to arbitrary member $x \in A$ can make up "set similar to x". Certainly this set shall be fuzzy one.

Example 4.14 Let's have a look at the member d in the fuzzy similarity relation which is previously introduced. Elements making similarity relationn with d shall be a, b and d itself. If these members are expressed according to the degree of similarity with d, we can get a similarity class of d.

$$\{(a, 0.4), (b, 0.4), (d, 1)\}$$

In the same manner, marking the elements similar to element e as a fuzzy set, the similarity class yields.

$$\{(c, 1), (e, 1), (f, 0.5)\} \quad \square$$

4.3.2 Fuzzy Compatibility Relation

Definition (Fuzzy compatibility relation) If fuzzy relation R in set A satisfies the following conditions, we call it "fuzzy compatibility relation" or "resemblance relation".

i) Reflexive relation

$$x \in A \Rightarrow \mu_R(x, x) = 1$$

ii) Symmetric relation

$$\forall (x, y) \in A \times A$$
$$\mu_R(x, y) = \mu \Rightarrow \mu_R(y, x) = \mu \quad \square$$

If fuzzy compatibility relation is given on set A, a partition can be processed into several subsets. Subsets from this partition are called the "fuzzy compatibility classes" and if we apply α-cut to the fuzzy compatibility relation, we get α-cut crisp compatibility relation R_α. A compatibility class A_i in this relation is defined by,

$$\mu_{R_\alpha}(x, y) = 1 \quad \text{if} \quad \mu_R(x, y) \geq \alpha \quad \forall x, y \in A_i$$
$$= 0 \quad \text{otherwise}$$

the collection of all compatibility classes from a α-cut is called complete α-cover. Note the differences of the cover and partition.

Example 4.15 (Fig 4.24.) is a description of covers made by α-cut and expression by undirected graph. Here by the help of $\alpha = 0.7$ cut, we get compatibility class $\{a, b\}$, $\{c, d, e\}$, $\{d, e, f\}$ and these compatibility classes cover the set A. Note that elements d and e are far from partition since they appear in dual subsets. The sequence of such compatibility classes constructs a compatibility covering tree as shown in (Fig 4.25)

	a	b	c	d	e	f
a	1.0	0.8	0.0	0.0	0.0	0.0
b	0.8	1.0	0.0	0.0	0.0	0.0
c	0.0	0.0	1.0	1.0	0.8	0.0
d	0.0	0.0	1.0	1.0	0.8	0.7
e	0.0	0.0	0.8	0.8	1.0	0.7
f	0.0	0.0	0.0	0.7	0.7	1.0

Fig. 4.24. Compatibility relation graph

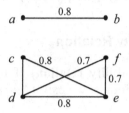

Fig. 4.24. (cont')

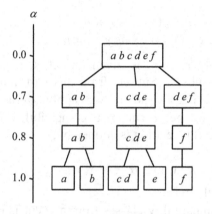

Fig. 4.25. Compatibility covering tree

4.3.3 Fuzzy Pre-order Relation

Definition (Fuzzy pre-order relation) Given fuzzy relation R in set A, if the followings are well kept for all $x, y, z \in A$, this relation is called pre-order relation.

i) Reflexive relation
$$\forall x \in A \Rightarrow \mu_R(x, x) = 1$$

ii) Transitive relation
$$\forall (x, y), (y, z), (x, z) \in A \times A$$
$$\mu_R(x, z) \geq \underset{y}{\text{Max}}[\text{Min}(\mu_R(x, y), \mu_R(y, z))] \quad \square$$

Also if certain relation is transitive but not reflexive, this relation is called "semi-pre-order" or "nonreflexive fuzzy pre-order".

Example 4.16 Here goes a semi-pre-order relation.

	a	b	c
a	0.2	1.0	0.4
b	0.0	0.6	0.3
c	0.0	1.0	0.3

□

If the membership function follows the relation $\mu_R(x, x) = 0$ for all x, we use the term "anti-reflexive fuzzy pre-order".

Example 4.17 The following matrix shows an anti-reflexive fuzzy pre-order relation.

	a	b	c
a	0.0	1.0	0.4
b	0.0	0.0	0.3
c	0.0	1.0	0.0

□

4.3.4 Fuzzy Order Relation

Definition (Fuzzy order relation) If relation R satisfies the followings for all $x, y, z \in A$, it is called fuzzy order relation.
  i) Reflexive relation
$$\forall x \in A \Rightarrow \mu_R(x, x) = 1$$

  ii) Antisymmetric relation
$$\forall (x, y) \in A \times A$$
$$\mu_R(x, y) \neq \mu_R(y, x) \text{ or } \mu_R(x, y) = \mu_R(y, x) = 0$$

  iii) Transitive relation
$$\forall (x, y), (y, z), (x, z) \in A \times A$$
$$\mu_R(x, z) \geq \underset{y}{\text{Max}}[\text{Min}(\mu_R(x, y), \mu_R(y, z))] \quad □$$

Example 4.18 For example, the relation shown in (Fig 4.26.) is a fuzzy order relation.

Definition (Corresponding crisp order) We can get a corresponding crisp relation R_1 from given fuzzy order relation R by arranging the value of membership function as follows.

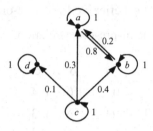

Fig. 4.26. Fuzzy order relation

i) if $\mu_R(x, y) \geq \mu_R(y, x)$ then $\mu_{R_1}(x,y)=1$ $\mu_{R_1}(y,x)=0$

ii) if $\mu R(x, y) = \mu R(y, x)$ then $\mu_{R_1}(x,y) = \mu_{R_1}(y,x) = 0$ □

Example 4.19 There is the corresponding crisp relation obtained from fuzzy order relation introduced in the previous example (Fig 4.27).

	a	b	c	d
a	1	1	0	0
b	0	1	0	0
c	1	1	1	1
d	0	0	0	1

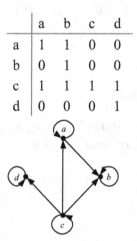

Fig. 4.27. Crisp order relation obtained from fuzzy order relation

If the corresponding order relation of a fuzzy order relation is total order or linear order, this fuzzy relation is named as "fuzzy total order", and if not, it is called "fuzzy partial order".

When the second condition of the fuzzy order relation is transformed to "perfect antisymmetric", the fuzzy order relation becomes a perfect fuzzy order.

(2') Perfect antisymmetric

$$\forall (x, y) \in A \times A, \quad x \neq y$$
$$\mu_R(x, y) > 0 \Rightarrow \mu_R(y, x) = 0$$

When the first condition (reflexivity) does not exist, the fuzzy order relation is called "fuzzy strict order".

Definition (Dominating and dominated class) In the fuzzy order relation, if $R(x, y) > 0$ holds, let us say that x dominates y and denote $x \geq y$. With this concept, two fuzzy sets are associated.

    i) The one is dominating class of element x. Dominating class $R_{\geq[x]}$ which dominates x is defined as,

$$\mu_{R_{\geq[x]}}(y) = \mu_R(y, x)$$

    ii) The other is dominated class. Dominated class $R_{\leq[x]}$ with elements dominated by x is defined as,

$$\mu_{R_{\leq[x]}}(y) = \mu_R(x, y) \quad \square$$

Example 4.20 Let's consider the following relation in (Fig 4.28.) Dominating class of element a and b shall be,

$$R_{\geq[a]} = \{(a, 1.0), (b, 0.7), (d, 1.0)\}$$
$$R_{\geq[b]} = \{(b, 1.0), (d, 0.9)\}$$

and dominated class by a is,

$$R_{\leq[a]} = \{(a, 1.0), (c, 0.5)\}$$

	a	b	c	d
a	1.0	0.0	0.5	0.0
b	0.7	1.0	0.7	0.0
c	0.0	0.0	1.0	0.0
d	1.0	0.9	1.0	1.0

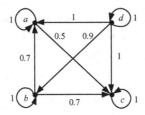

Fig. 4.28. Example for dominating relation

If such fuzzy order relation is given, fuzzy upper bound of subset $A' = \{x, y\}$ can be obtained by fuzzy intersection or Min operation of the dominating class.

$$\bigcap_{x \in A'} R_{\geq[x]}$$

Example 4.21 If $A' = \{a, b\}$ in the above example, the fuzzy upper bound shall be.

$R_{\geq[a]} \cap R_{\geq[b]} = \{(a, 1.0), (b, 0.7), (d, 1.0)\} \cap \{(b, 1.0), (d, 0.9)\} = \{(b, 0.7), (d, 0.9)\}$

In the fuzzy set describing the upper bound of $\{a, b\}$, since b is itself, d may well be the only least upper bound.   □

4.4 Other Fuzzy Relations

4.4.1 Fuzzy Ordinal Relation

Definition (Fuzzy ordinal relation) If fuzzy relation R is defined in A and the following is appreciated for all $x, y \in A$, its corresponding crisp order relation is defined as fuzzy ordinal relation.

i) Reflexive relation

$$x \in A \Rightarrow \mu_R(x, x) = 1$$

ii) Antisymmetric relation

$$\forall (x, y) \in A \times A$$
$$\mu_R(x, y) \neq \mu_R(y, x) \text{ or } \mu_R(x, y) = \mu_R(y, x) = 0$$

iii) Let graph G_R represent the corresponding crisp order relation obtained by arranging the membership as in sec 4.3.4.
- if $\mu_R(x, y) > \mu_R(y, x)$ then $\mu_{G_R}(x, y) = 1$

$$\mu_{G_R}(y, x) = 0$$

- if $\mu_R(x, y) = \mu_R(y, x)$ then $\mu_{G_R}(x, y) = \mu_{G_R}(y, x) = 0$

then, such obtained graph G_R does not bear any circuit.   □

Example 4.22 In fuzzy relation R in (Fig 4.29.), we see that there exist circuits. From this relation, if we arrange the membership degrees by the above procedure, the circuits were disappeared as shown in (Fig 4.30.)   □

	a	b	c	d	e
a	1.0	0.9	0.0	0.0	0.8
b	0.4	1.0	0.5	1.0	0.3
c	0.9	0.8	1.0	0.0	0.0
d	0.0	0.0	0.0	1.0	0.0
e	0.2	0.0	0.0	0.0	1.0

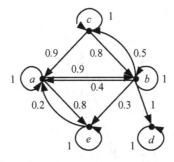

Fig. 4.29. Fuzzy graph

	a	b	c	d	e
a	1	1	0	0	1
b	0	1	0	1	1
c	1	1	1	0	0
d	0	0	0	1	0
e	0	0	0	0	1

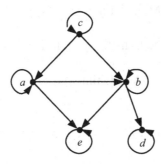

Fig. 4.30. Graph corresponding to fuzzy graph in (Fig 4.29.)

Example 4.23 Applying the ordinal function described in section 3.2.5 to the relation R in the previous example, we get an order like in (Fig 4.31.)

$$f(d) = f(e) = 0$$
$$f(b) = 1$$
$$f(a) = 2$$
$$f(c) = 3 \quad \square$$

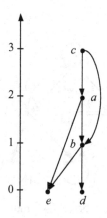

Fig. 4.31. Ordinal function

4.4.2 Dissimilitude Relation

We have seen the three conditions for the similarity relation.
(1)   Reflexive relation
(2)   Symmetric relation
(3)   Transitive relation(Max - Min transitivity)

Especially, we write the transitivity as follows, for $(x, y), (y, z), (x, z) \in A \times A$

$$\mu_R(x, z) \geq \underset{y}{\text{Max}}[\text{Min}[\mu_R(x, y), \mu_R(y, z)]]$$

or

$$\mu_R(x, z) \geq \underset{y}{\vee}[\mu_R(x, y) \wedge \mu_R(y, z)]$$

Dissimilitude relation maintains the opposite position in the concept of similarity relation. As a result applying the complement relation $\overline{R}$ instead of relation R, we can think of the transitivity of $\overline{R}$.

For any $(x, y) \in A \times A$, since $\mu_{\bar{R}}(x,y) = 1 - \mu_R(x,y)$, transitivity of R shall be,

$$\mu_R(x,z) \geq \vee_{y} [(1 - \mu_{\bar{R}}(x,y)) \wedge (1 - \mu_{\bar{R}}(y,z))]$$

the right part of this relation can be transformed by $\overline{A \cap B} = \overline{A} \cup \overline{B}$,

$$1 - (\mu_{\bar{R}}(x,y)) \wedge (1 - \mu_{\bar{R}}(y,z)) = 1 - (\mu_R(x,y) \vee \mu_R(y,z))$$

consequently,

$$\mu_R(x,z) \geq \vee_{y} [1 - (\mu_R(x,y) \vee \mu_R(y,z))]$$

$$\mu_R(x,z) \leq \wedge_{y} [\mu_R(x,y) \vee \mu_R(y,z)]$$

so this property is called transitivity of Min - Max operation.

Definition (Dissimilitude relation) We can define dissimilitude relation for fuzzy relation $R \subseteq A \times A$,

 i) Antiflexive relation

$$x \in A \Rightarrow \mu_R(x, x) = 0$$

 ii) Symmetric relation

for $\forall\ (x, y) \in A \times A, \quad \mu_R(x, y) = \mu_R(y, x)$

iii) Min - Max transitive relation

for $\forall\ (x, y), (y, z), (x, z) \in A \times A \quad \mu_R(x, z) \leq \wedge_{y} [\mu_R(x, y) \vee \mu_R(y, z)]$ □

Example 4.24 Let's consider a relation R in the following.

	a	b	c	d
a	0.0	0.2	0.3	0.0
b	0.2	0.0	0.3	0.2
c	0.3	0.3	0.0	0.3
d	0.0	0.2	0.3	0.0

In this relation R, we can easily see it is an antiflexive and symmetric relation. To understand the transitive relation of Min - Max operation, investigate the pair of (a, b).

First, for $\mu(a, b) = 0.2$, we check paths with length 2

$$\mu(a, a) \vee \mu(a, b) = 0.0 \vee 0.2 = 0.2$$
$$\mu(a, b) \vee \mu(b, b) = 0.2 \vee 0.0 = 0.2$$
$$\mu(a, c) \vee \mu(c, b) = 0.3 \vee 0.3 = 0.3$$
$$\mu(a, d) \vee \mu(d, b) = 0.0 \vee 0.2 = 0.2$$

the minimum of these values is 0.2, consequently $\mu(a, b) \leq 0.2$ for the pair (a, b), the Min-Max transitivity is maintained. In the same manner, we can see, for all pairs, this transitivity holds.   □

4.4.3 Fuzzy Morphism

Definition (Homomorphism) Given multiple crisp relations $R \subseteq A \times A$ and $S \subseteq B \times B$, homomorphism from (A, R) to (B, S) is for the function $h : A \to B$ having the characteristics as,

for $x_1, x_2 \in A$

$$(x_1, x_2) \in R \Rightarrow (h(x_1), h(x_2)) \in S$$

In other words, if two elements x_1 and x_2 are related by R, their images $h(x_1)$ and $h(x_2)$ are also related by S.   □

Definition (Strong homomorphism) Given two crisp relations $R \subseteq A \times A$ and $S \subseteq B \times B$, if the function $h : A \to B$ satisfies the followings, it is called strong homomorphism from (A, R) to (B, S).

i) For all $x_1, x_2 \in A$,

$$(x_1, x_2) \in R \Rightarrow (h(x_1), h(x_2)) \in S$$

ii) For all $y_1, y_2 \in B$,

if $x_1 \in h^{-1}(y_1), x_2 \in h^{-1}(y_2)$ then $(y_1, y_2) \in S \Rightarrow (x_1, x_2) \in R$

In other words, the inverse image $(x_1, x_2) \in R$ of $(y_1, y_2) \in S$ always stands for the homomorphism related by R. Here h, we see, is a many-to-one mapping function.   □

Definition (Fuzzy homomorphism) If the relations $R \subseteq A \times A$ and $S \subseteq B \times B$ are fuzzy relations, the above morphism is extended to a fuzzy homomorphism as follows.

For all $x_1, x_2 \in A$ and their images $h(x_1), h(x_2) \in B$,

$$\mu_R(x_1, x_2) \leq \mu_S[h(x_1), h(x_2)]$$

in other words, the strength of the relation S for $(h(x_1), h(x_2))$ is stronger than or equal to the that of R for (x_1, x_2).   □

If a homomophism exists between fuzzy relations (A, R) and (B, S), the homomorphism h partitions A into subsets $A_1, A_2, \dots A_n$ because it is a many-to-one mapping

$$\forall x_i \in A_j \; , \quad i = 1, 2, \dots n \qquad\qquad h(x_i) = y \in B$$

so to speak, image $h(x_i)$ of elements x_i in A_j is identical to element y in B. In this manner, every element in A shall be mapped to one of B. If the strength between A_j and A_k gets the maximum strength between $x_j \in A_j$, and $x_k \in A_k$, this morphism is replaced with fuzzy strong homomorphism.

Definition (Fuzzy strong homomorphism) Given the fuzzy relations R and S, if h satisfies the followings, h is a fuzzy strong homomorphism.

For all $x_j \in A_j$, $x_k \in A_k$, A_j, $A_k \subseteq A$
$$y_1 = h(x_j), \quad y_2 = h(x_k)$$
$$y_1, y_2 \in B, (y_1, y_2) \in S,$$
$$\underset{x_j, x_k}{\text{Max}} \ \mu_R(x_j, x_k) = \mu_S(y_1, y_2) \quad \square$$

4.4.4 Examples of Fuzzy Morphism

Example 4.25 Consider the relations $R \subseteq A \times A$ and $S \subseteq B \times B$ in the following

R	a	b	c	d
a	0.0	0.6	0.0	0.0
b	0.0	0.0	0.8	0.0
c	1.0	0.0	0.0	0.0
d	0.0	0.6	0.0	0.0

S	α	β	γ
α	0.6	0.8	0.0
β	1.0	0.0	0.6
γ	0.6	0.0	0.0

we apply the mapping function h from A to B as follows.
$$h : a, b \ \rightarrow \ \alpha$$
$$c \ \rightarrow \ \beta$$
$$d \ \rightarrow \ \gamma$$

Here all $(x_1, x_2) \in R$ of A has the relation $(h(x_1), h(x_2)) \in S$ in B. Furthermore $\mu_R(x_1, x_2) \le \mu_S(h(x_1), h(x_2))$ holds. For example, $h(c) = \beta$, $h(d) = \gamma$, $\mu_R(c, d) = 0 \le \mu_S(\beta, \gamma) = 0.6$

As a consequence, this morphism is a fuzzy homomorphism (Fig 4.32). We know $\mu_S(\beta, \gamma) = 0.6$, but we are not able to find its corresponding pair in R. Therefore it is not a fuzzy strong homomorphism. $\square$

Example 4.26 We have two relations $R \subseteq A \times A$ and $S \subseteq B \times B$.

R	a	b	c	d	e
a	0.5	0.5	0.0	0.0	0.0
b	1.0	0.0	0.5	0.0	0.0
c	0.0	0.0	0.0	1.0	0.5
d	0.0	0.0	0.9	0.0	0.0
e	0.0	0.0	0.0	1.0	0.0

S	α	β	γ
α	0.5	0.5	0.0
β	1.0	0.5	1.0
γ	0.0	0.9	1.0

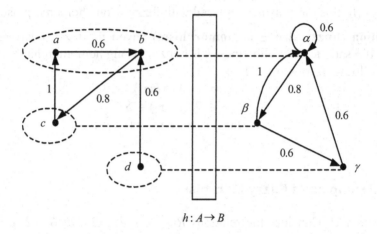

$$h : A \rightarrow B$$

Fig. 4.32. Fuzzy homomorphism

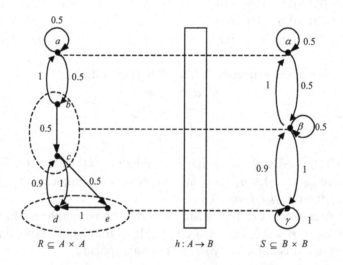

$R \subseteq A \times A$ $h : A \rightarrow B$ $S \subseteq B \times B$

Fig. 4.33. Fuzzy strong homomorphism

We have also a function from A to B as follows.

$$
\begin{array}{ccc}
h : a & \rightarrow & \alpha \\
b, c & \rightarrow & \beta \\
d, e & \rightarrow & \gamma
\end{array}
$$

Here for all $(x_1, x_2) \in R$, $(h(x_1), h(x_2)) \in S$ in B and inversely, for all $(y_1, y_2) \in S$, $(h^{-1}(y_1), h^{-1}(y_2)) \in R$ in A. Thus h completes the conditions for fuzzy homomorphism. Now, let's consider the conditions for the fuzzy strong homomorphism.

For example, $(\beta, \gamma) \in S$, $\mu_S(\beta, \gamma) = 1$,
$$h^{-1}(\beta) = \{b, c\}, h^{-1}(\gamma) = \{d, e\}$$

$$\text{Max} [\mu_R(c, d), \mu_R(c, e)] = \text{Max} [1, 0.5] = 1 = \mu_S(\beta, \gamma)$$
in the same manner, we can verify for other pairs and then we see the morphism h is a fuzzy strong homomorphism (Fig 4.33).   □

[SUMMARY]

☐ Fuzzy graph
  - $G = (V, E)$
  - V : set of elements, nodes or vertices.
  - E : set of pairs of nodes.

☐ A fuzzy graph represents a fuzzy relation

☐ α-cut of a fuzzy relation

$$\alpha_1 \geq \alpha_2 \implies R_{\alpha_1} \subseteq R_{\alpha_2}$$
$$G_{\alpha_1} \subseteq G_{\alpha_2}$$

☐ Characteristics of a fuzzy relation
  - Assume R is a fuzzy relation defined on A,
  - Reflexive relation
       if, for all $x \in A$, $\mu_R(x, x) = 1$, then R is said to be ***reflexive***.
  - Symmetric relation
       if $\forall \, (x, y) \in A \times A, \quad \mu_R(x, y) = \mu \implies \mu_R(y, x) = \mu,$
       then R is said to be ***symmetric***.

       if $\forall \, (x, y) \in A \times A$ with $x \neq y$, $\mu_R(x, y) \neq \mu_R(y, x)$
                              or $\mu_R(x, y) = \mu_R(y, x) = 0,$
       then R is said to be ***antisymmetric***

       if $\exists \, (x, y) \in A \times A$ with $x \neq y$, *such that* $\mu_R(x, y) \neq \mu_R(y, x)$,
       then R is said to be ***asymmetric***.

       if $\forall \, (x, y) \in A \times A$ with $x \neq y$, $\mu_R(x, y) > 0 \implies \mu_R(y, x) = 0,$
       then R is said to be ***perfectly antisymmetric***.

  - Transitive relation
       if $\forall \, (x, y), (y, z), (x, z) \in A \times A \ \mu_R(x, z) \geq \underset{y}{\text{Max}}[\text{Min}(\mu_R(x, y)$
       $\mu_R(y, z))]$
       then R is said to be ***transitive***.

☐ Fuzzy equivalence relation, similarity relation
  - Reflexive relation
  - Symmetric relation

  - Transitive relation
  - Partition, similarity class

☐ Fuzzy compatibility relation
  - Reflexive relation
  - Symmetric relation
  - Fuzzy compatibility class

☐ Fuzzy pre-order relation
  - Reflexive relation
  - Transitive relation

☐ Fuzzy order relation
  - Reflexive relation
  - Symmetric relation
  - Transitive relation
  - Total order, partial order
  - Fuzzy strict order, perfect fuzzy order
  - Dominating class, dominated class

☐ Fuzzy ordinal relation
  - Reflexive relation
  - Antisymmetric relation

☐ Dissimilitude relation
  - Reflexive relation
  - Symmetric relation
  - Transitive relation (Max-Min transitivity)

☐ Fuzzy homomorphism : h
$$\mu_R(x_1, x_2) \leq \mu_S[h(x_1), h(x_2)]$$
$$x_1, x_2 \in A$$
$$h(x_1), h(x_2) \in B$$

☐ Fuzzy strong homomorphism : h
$$x_1, x_2 \in A, y_1 = h(x_1), y_2 = h(x_2)$$
$$y_1, y_2 \in B, (y_1, y_2) \in S$$
$$\underset{x_1, x_2}{\text{Max}} \quad \mu_R(x_1, x_2) = \mu_S(y_1, y_2)$$

[EXERCISES]

4.1 There is a fuzzy relation $R = \{(x, y) \mid y \approx x^2, x \in \mathcal{R}, y \in \mathcal{R}\}$. The symbol $\approx$ means "close to". Show the graphical representation of this relation.

4.2 Let R be a fuzzy relation defined on $\mathcal{R}^2$, where $x \in \mathcal{R}^+$ and $y \in \mathcal{R}^+$,
$$\mu_R(x, y) = 1 - |x - y| \quad \text{if} \quad 0 \le |x - y| \le 1$$
$$= 0 \quad \text{otherwise}$$
Show the fuzzy graph representing the relation.

4.3 Consider two fuzzy relations R and S defined on $\mathcal{R}^2$ such that, for x, y $\in \mathcal{R}$,

R represents "$|y - x|$ close to α"
S represents "$|y - x|$ close to β"

Show the graphical form of R and S.

4.4 Show the graphical representations of the following relations.
a) $\mu_R(x, y) = e^{-(x-y)^2}, \quad x, y \in \mathcal{R}$
b) $\mu_R(x, y) = e^{-k(x-y)^2}, \quad k \ge 1, \quad x, y \in \mathcal{R}$

4.5 Consider two fuzzy sets A and B defined in the real numbers.

$A = \{x \mid x \text{ is close to } 2\pi\}$
$\mu_B(y) = \cos x, x \in A$

Show the graphical representations of A and B.

4.6 Show the graphical representation of the relation R
$$\mu_R(x, y) = 1 - (x^2 + y^2)^{1/2} \le 1 \quad \text{where } x \in \mathcal{R} \text{ and } y \in \mathcal{R}.$$

4.7 Determine whether the following fuzzy relation is an equivalence relation.

	a	b	c	d
a	1.0	0.8	0.4	0.1
b	0.8	1.0	0.0	0.0
c	0.4	0.0	1.0	0.5
d	0.1	0.0	0.5	1.0

4.8 Discuss the properties of the following morphism h.

$$
\begin{aligned}
h : a, b &\rightarrow \alpha \\
c &\rightarrow \beta \\
d, e &\rightarrow \gamma
\end{aligned}
$$

where $A = \{a, b, c, d, e\}$,
            $B = \{\alpha, \beta, \gamma\}$,
            $R \subseteq A \times A$,
            $S \subseteq B \times B$

R	a	b	c	d	e
a	1.0	0.0	0.0	1.0	0.1
b	1.0	0.4	0.0	0.0	1.0
c	0.4	0.0	0.0	0.9	0.0
d	0.8	0.0	0.6	0.8	0.0
e	0.0	0.6	0.8	0.4	1.0

S	α	β	γ
α	1.0	0.0	1.0
β	0.4	0.0	0.9
γ	0.8	0.8	1.0

4.9 Consider a fuzzy relation R such that

$$
\mu_R = \frac{1}{1 + x^2 + y^2}
$$

a) Show the graphical form of this relation
b) Determine the α-cut relation for $\alpha = 0.3$
c) Determine the sub-relation of R with $\mu_R \geq 0.5$.

4.10. Determine whether the following relations $R \subseteq A \times A$ is an equivalence relation and construct a partition tree of the set $A = \{a, b, c, d, e\}$

	a	b	c	d	e
a	1.0	0.7	0	0.4	0
b	0.7	1.0	0	0.4	0
c	0	0	1.0	0	1.0
d	0.4	0.4	0	1.0	0
e	0	0	1.0	0	1.0

4.11. Consider a perfect antisymmetric relation R. Determine the following statement is true or not : $R \cap R^{-1} = \emptyset$
a) When R is a crisp relation

b) When R is a fuzzy relation

4.12. Show the following statement are true or not.

a) When fuzzy relation $R \subseteq A \times A$ is Max-Min transitive, then
$$R^2 \subseteq R$$
b) When fuzzy relation $R \subseteq A \times A$ is Min-Max transitive, then
$$R \subseteq R^2$$

4.13 Consider a fuzzy relation $R \subseteq A \times A$ where $A = \{a, b, c\}$.

R	a	b	c
a	1.0	0.8	0.7
b	0.8	1.0	0.7
c	0.7	0.7	1.0

a) Show the relation is an equivalence relatio
b) Determine the partition tree of the set $A = \{a, b, c\}$ by using R

4.14 There is a set $A \subset \mathcal{R}$ where $\mathcal{R}$ is the set of real numbers and $x \in \mathcal{R}$.
$A = \{x \mid x$ is close to $(4k +1)\pi/2, k=\cdots-2, -1, 0, 1, 2\cdots \}$

a) Show the graphical representation of the set A.
b) Show the graphical representation relation R defined as follows
$$\mu_R(x, y) = sin\ x \quad \text{where } x \in A, y \in \mathcal{R}$$
c) Show the graphical representation of set B defined as follows
$$\mu_B(y) = sin\ x \quad \text{where} \quad x \in A, y \in \mathcal{R}$$

Chapter 5. FUZZY NUMBER

This chapter describes fuzzy numbers. First of all, we'll look into interval, the fundamental concept of fuzzy number, and then operation of fuzzy numbers. In addition, we'll introduce special kind of fuzzy number such as triangular fuzzy number and trapezoidal fuzzy number.

5.1 Concept of Fuzzy Number

5.1.1 Interval

When interval is defined on real number $\Re$, this interval is said to be a subset of $\Re$. For instance, if interval is denoted as $A = [a_1, a_3]$ $a_1, a_3 \in \Re$, $a_1 < a_3$, we may regard this as one kind of sets. Expressing the interval as membership function is shown in the following (Fig 5.1) :

$$\mu_A(x) = \begin{cases} 0, & x < a_1 \\ 1, & a_1 \leq x \leq a_3 \\ 0, & x > a_3 \end{cases}$$

If $a_1 = a_3$, this interval indicates a point. That is, $[a_1, a_1] = a_1$

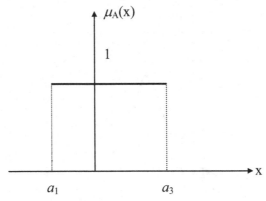

Fig. 5.1. Interval $A = [a_1, a_3]$

5.1.2 Fuzzy Number

Fuzzy number is expressed as a fuzzy set defining a fuzzy interval in the real number $\Re$. Since the boundary of this interval is ambiguous, the interval is also a fuzzy set. Generally a fuzzy interval is represented by two end points a_1 and a_3 and a peak point a_2 as $[a_1, a_2, a_3]$ (Fig 5.2). The a-cut operation can be also applied to the fuzzy number. If we denote a-cut interval for fuzzy number A as A_α, the obtained interval A_α is defined as

$$A_\alpha = [a_1^{(\alpha)}, a_3^{(\alpha)}]$$

We can also know that it is an ordinary crisp interval (Fig 5.3). We review here the definition of fuzzy number given in section 1.5.4.

Definition (Fuzzy number) It is a fuzzy set the following conditions :
  – convex fuzzy set
  – normalized fuzzy set
  – it's membership function is piecewise continuous.
  – It is defined in the real number.   □

Fuzzy number should be normalized and convex. Here the condition of normalization implies that maximum membership value is 1.

$$\exists x \in \Re, \qquad \mu_A(x) = 1$$

The convex condition is that the line by α-cut is continuous and α-cut interval satisfies the following relation.

$$A_\alpha = [a_1^{(\alpha)}, a_3^{(\alpha)}]$$

$$(\alpha' < \alpha) \Rightarrow (a_1^{(\alpha')} \leq a_1^{(\alpha)}, a_3^{(\alpha')} \geq a_3^{(\alpha)})$$

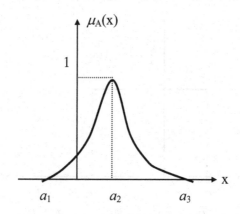

Fig. 5.2. Fuzzy Number $A = [a_1, a_2, a_3]$

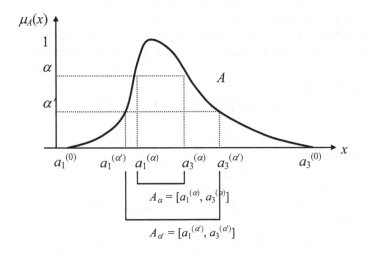

Fig. 5.3. α-cut of fuzzy number $(\alpha' < \alpha) \Rightarrow (A_\alpha \subset A_{\alpha'})$

The convex condition may also be written as,
$$(\alpha' < \alpha) \Rightarrow (A_\alpha \subset A_{\alpha'})$$

5.1.3 Operation of Interval

Operation of fuzzy number can be generalized from that of crisp interval.
Let's have a look at the operations of interval.
$\forall a_1, a_3, b_1, b_3 \in \mathfrak{R}$
$$A = [a_1, a_3], B = [b_1, b_3]$$
Assuming A and B as numbers expressed as interval, main operations of
interval are
(1)   Addition
$$[a_1, a_3] \, (+) \, [b_1, b_3] = [a_1 + b_1, a_3 + b_3]$$
(2)   Subtraction
$$[a_1, a_3] \, (-) \, [b_1, b_3] = [a_1 - b_3, a_3 - b_1]$$
(3)   Multiplication
$[a_1, a_3] \, (\bullet) \, [b_1, b_3] = [a_1 \bullet b_1 \wedge a_1 \bullet b_3 \wedge a_3 \bullet b_1 \wedge a_3 \bullet b_3, a_1 \bullet b_1 \vee a_1 \bullet b_3$
$\vee a_3 \bullet b_1 \vee a_3 \bullet b_3]$
(4)   Division
$[a_1, a_3] \, (/) \, [b_1, b_3] = [a_1 / b_1 \wedge a_1 / b_3 \wedge a_3 / b_1 \wedge a_3 / b_3, a_1 / b_1 \vee a_1 / b_3 \vee$
$a_3 / b_1 \vee a_3 / b_3]$
excluding the case $b_1 = 0$ or $b_3 = 0$
(5)   Inverse interval

$$[a_1, a_3]^{-1} = [1 / a_1 \wedge 1 / a_3, 1 / a_1 \vee 1 / a_3]$$

excluding the case $a_1 = 0$ or $a_3 = 0$

When previous sets A and B is defined in the positive real number $\Re+$, the operations of multiplication, division, and inverse interval are written as,

(3')   Multiplication

$$[a_1, a_3] (\bullet) [b_1, b_3] = [a_1 \bullet b_1, a_3 \bullet b_3]$$

(4')   Division

$$[a_1, a_3] (/) [b_1, b_3] = [a_1 / b_3, a_3 / b_1]$$

(5')   Inverse Interval

$$[a_1, a_3]^{-1} = [1 / a_3, 1 / a_1]$$

(6)   Minimum

$$[a_1, a_3] (\wedge) [b_1, b_3] = [a_1 \wedge b_1, a_3 \wedge b_3]$$

(7)   Maximum

$$[a_1, a_3] (\vee) [b_1, b_3] = [a_1 \vee b_1, a_3 \vee b_3]$$

Example 5.1 There are two intervals A and B,

$$A = [3, 5], B = [-2, 7]$$

then following operation might be set.

$$A(+)B \quad = \quad [3-2, \quad 5+7] \quad = \quad [1, \quad 12]$$

$$A(-)B \quad = \quad [3-7, \quad 5-(-2)] \quad = \quad [-4, \quad 7]$$

$$A(\bullet)B \quad = \quad [3\bullet(-2)\wedge3\bullet7\wedge5\bullet(-2)\wedge5\bullet7, \quad 3\bullet(-2)\vee\cdots]$$
$$= \quad [-10, \quad 35]$$

$$A(/)B \quad = \quad [3/(-2)\wedge3/7\wedge5/(-2)\wedge5/7, \quad 3/(-2)\vee\cdots]$$
$$= \quad [-2.5, \quad 5/7]$$

$$B^{-1} = [-2,7]^{-1} = \left[\frac{1}{(-2)} \wedge \frac{1}{7}, \quad \frac{1}{(-2)} \vee \frac{1}{7} \right] = \left[-\frac{1}{2}, \quad \frac{1}{7} \right] \qquad \square$$

5.2 Operation of Fuzzy Number

5.2.1 Operation of α-cut Interval

We referred to α-cut interval of fuzzy number $A = [a_1, a_3]$ as crisp set

$$A_\alpha = [a_1^{(\alpha)}, a_3^{(\alpha)}], \forall \alpha \in [0, 1], a_1, a_3, a_1^{(\alpha)}, a_3^{(\alpha)} \in \Re$$

so A_α is a crisp interval. As a result, the operations of interval reviewed in the previous section can be applied to the α-cut interval A_α.

If α-cut interval B_α of fuzzy number B is given

$$B = [b_1, b_3], \quad b_1, b_3, \in \Re$$
$$B_\alpha = [b_1^{(\alpha)}, b_3^{(\alpha)}], \forall \alpha \in [0, 1], b_1^{(\alpha)}, b_3^{(\alpha)} \in \Re,$$

operations between A_α and B_α can be described as follows :

$$[a_1^{(\alpha)}, a_3^{(\alpha)}] \; (+) \; [b_1^{(\alpha)}, b_3^{(\alpha)}] = [a_1^{(\alpha)} + b_1^{(\alpha)}, a_3^{(\alpha)} + b_3^{(\alpha)}]$$
$$[a_1^{(\alpha)}, a_3^{(\alpha)}] \; (-) \; [b_1^{(\alpha)}, b_3^{(\alpha)}] = [a_1^{(\alpha)} - b_3^{(\alpha)}, a_3^{(\alpha)} - b_1^{(\alpha)}]$$

these operations can be also applicable to multiplication and division in the same manner.

5.2.2 Operation of Fuzzy Number

Previous operations of interval are also applicable to fuzzy number. Since outcome of fuzzy number (fuzzy set) is in the shape of fuzzy set, the result is expressed in membership function.

$\forall x, y, z \in \Re$

(1) Addition: $A \; (+) \; B$

$$\mu_{A(+)B}(z) = \bigvee_{z=x+y} (\mu_A(x) \wedge \mu_B(y))$$

(2) Subtraction: $A \; (-) \; B$

$$\mu_{A(-)B}(z) = \bigvee_{z=x-y} (\mu_A(x) \wedge \mu_B(y))$$

(3) Multiplication: $A \; (\bullet) \; B$

$$\mu_{A(\bullet)B}(z) = \bigvee_{z=x\bullet y} (\mu_A(x) \wedge \mu_B(y))$$

(4) Division: $A \; (/) \; B$

$$\mu_{A(/)B}(z) = \bigvee_{z=x/y} (\mu_A(x) \wedge \mu_B(y))$$

(5) Minimum: $A \; (\wedge) \; B$

$$\mu_{A(\wedge)B}(z) = \bigvee_{z=x\wedge y} (\mu_A(x) \wedge \mu_B(y))$$

(6) Maximum: $A \; (\vee) \; B$

$$\mu_{A(\vee)B}(z) = \bigvee_{z=x\vee y} (\mu_A(x) \wedge \mu_B(y))$$

We can multiply a scalar value to the interval. For instance, multiplying $a \in \Re$,

$$a[b_1, b_3] = [a \bullet b_1 \wedge a \bullet b_3, a \bullet b_1 \vee a \bullet b_3]$$

Example 5.2 There is a scalar multiplication to interval. Note the scalar value is negative.

$$-4.15 \; [-3.55, 0.21] = [(-4.15) \bullet (-3.55) \wedge (-4.15) \bullet 0.21, (-4.15) \bullet (-3.55)$$
$$\vee (-4.15) \bullet 0.21]$$
$$= [14.73 \wedge -0.87, 14.73 \vee -0.87]$$
$$= [-0.87, 14.73] \quad \square$$

We can also multiply scalar value to α-cut interval of fuzzy number.

$\forall \alpha \in [0, 1], b_1^{(\alpha)}, b_3^{(\alpha)} \in \Re$

$$a[b_1^{(\alpha)}, b_3^{(\alpha)}] = [a \bullet b_1^{(\alpha)} \wedge a \bullet b_3^{(\alpha)}, a \bullet b_1^{(\alpha)} \vee a \bullet b_3^{(\alpha)}] \quad \square$$

5.2.3 Examples of Fuzzy Number Operation

Example 5.3 Addition A(+)B
For further understanding of fuzzy number operation, let us consider two fuzzy sets A and B. Note that these fuzzy sets are defined on discrete numbers for simplicity.
$$A = \{(2, 1), (3, 0.5)\}, B = \{(3, 1), (4, 0.5)\}$$
First of all, our concern is addition between A and B. To induce $A(+)B$, for all $x \in A$, $y \in B$, $z \in A(+)B$, we check each case as follows(Fig 5.4) :
  i)   for $z < 5$,
$$\mu_{A(+)B}(z) = 0$$
  ii) $z = 5$
$$\text{results from } x + y = 2 + 3$$
$$\mu_A(2) \wedge \mu_B(3) = 1 \wedge 1 = 1$$
$$\mu_{A(+)B}(5) = \bigvee_{5=2+3}(1) = 1$$
  iii) $z = 6$
$$\text{results from } x + y = 3 + 3 \text{ or } x + y = 2 + 4$$
$$\mu_A(3) \wedge \mu_B(3) = 0.5 \wedge 1 = 0.5$$
$$\mu_A(2) \wedge \mu_B(4) = 1 \wedge 0.5 = 0.5$$
$$\mu_{A(+)B}(6) = \bigvee_{\substack{6=3+3 \\ 6=2+4}}(0.5, 0.5) = 0.5$$
  iv) $z = 7$
$$\text{results from } x + y = 3 + 4$$
$$\mu_A(3) \wedge \mu_B(4) = 0.5 \wedge 0.5 = 0.5$$
$$\mu_{A(+)B}(7) = \bigvee_{7=3+4}(0.5) = 0.5$$
  v)   for $z > 7$
$$\mu_{A(+)B}(z) = 0$$
so $A(+)B$ can be written as
$$A(+)B = \{(5, 1), (6, 0.5), (7, 0.5)\} \quad \square$$

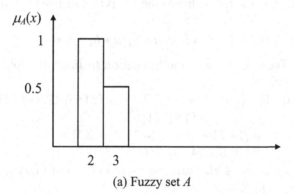

(a) Fuzzy set A

Fig. 5.4. Add operation of fuzzy set

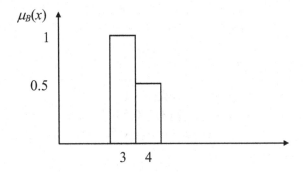

(b) Fuzzy number B

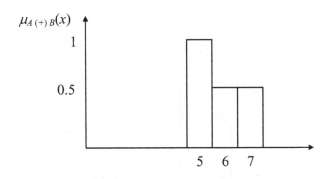

(c) Fuzzy set $A (+) B$

Fig. 5.4. (cont')

Example 5.5 Subtraction $A(-)B$
Let's manipulate $A(-)B$ between our previously defined fuzzy set A and B.
For $x \in A$, $y \in B$, $z \in A(-)B$, fuzzy set $A(-)B$ is defined as follows (Fig5.
5).

   i) For $z < -2$,

$$\mu_{A(-)B}(z) = 0$$

   ii) $z = -2$

$$\text{results from } x - y = 2 - 4$$
$$\mu_A(2) \wedge \mu_B(4) = 1 \wedge 0.5 = 0.5$$
$$\mu_{A(-)B}(-2) = 0.5$$

   iii) $z = -1$

$$\text{results from } x - y = 2 - 3 \text{ or } x - y = 3 - 4$$

$$\mu_A(2) \wedge \mu_B(3) = 1 \wedge 1 = 1$$
$$\mu_A(3) \wedge \mu_B(4) = 0.5 \wedge 0.5 = 0.5$$
$$\mu_{A(-)B}(-1) = \underset{\substack{-1=2-3 \\ -1=3-4}}{\vee}(1, 0.5) = 1$$

iv) $z = 0$

$$\text{results from } x - y = 3 - 3$$
$$\mu_A(3) \wedge \mu_B(3) = 0.5 \wedge 1 = 0.5$$
$$\mu_{A(-)B}(0) = 0.5$$

v) For $z \geq 1$

$$\mu_{A(-)B}(z) = 0$$
so $A(-)B$ is expressed as
$$A(-)B = \{(-2, 0.5), (-1, 1), (0, 0.5)\} \quad \square$$

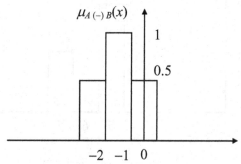

$\mu_{A\,(-)\,B}(x)$

Fig. 5.5. Fuzzy number $A\,(-)\,B$

Example 5.6 Max operation $A(\vee)B$
Let's deal with the operation Max $A(\vee)B$ between A and B
for $x \in A$, $y \in B$, $z \in A(\vee)B$, fuzzy set $A(\vee)B$ is defined by $\mu_{A(\vee)B}(z)$.
  i) $z \leq 2$

$$\mu_{A(\vee)B}(z) = 0$$

ii) $z = 3$
  from $x \vee y = 2 \vee 3$ and $x \vee y = 3 \vee 3$
$$\mu_A(2) \wedge \mu_B(3) = 1 \wedge 1 = 1$$
$$\mu_A(3) \wedge \mu_B(3) = 0.5 \wedge 1 = 0.5$$
$$\mu_{A(\vee)B}(3) = \underset{\substack{3=2\vee3 \\ 3=3\vee3}}{\vee}(1, 0.5) = 1$$

iii) $z = 4$
  from $x \vee y = 2 \vee 4$ and $x \vee y = 3 \vee 4$
$$\mu_A(2) \wedge \mu_B(4) = 1 \wedge 0.5 = 0.5$$

$$\mu_A(3) \wedge \mu_B(4) = 0.5 \wedge 0.5 = 0.5$$
$$\mu_{A(\vee)B}(4) = \underset{\substack{4=2\vee4 \\ 4=3\vee4}}{\vee}(0.5, 0.5) = 0.5$$

v) $z > 5$

impossible $\mu_{A(\vee)B}(z) = 0$

so $A(\vee)B$ is defined to be

$$A(\vee)B = \{(3, 1), (4, 0.5)\} \quad \square$$

so far we have seen the results of operations are fuzzy sets, and thus we come to realize that the extension principle is applied to the operation of fuzzy number.

5.3 Triangular Fuzzy Number

5.3.1 Definition of Triangular Fuzzy Number

Among the various shapes of fuzzy number, triangular fuzzy number(TFN) is the most popular one.

Definition (Triangular fuzzy number) It is a fuzzy number represented with three points as follows :

$$A = (a_1, a_2, a_3)$$

this representation is interpreted as membership functions (Fig5.6).

$$\mu_{(A)}(x) = \begin{cases} 0, & x < a_1 \\ \dfrac{x - a_1}{a_2 - a_1}, & a_1 \le x \le a_2 \\ \dfrac{a_3 - x}{a_3 - a_2}, & a_2 \le x \le a_3 \\ 0, & x > a_3 \end{cases} \quad \square$$

Now if you get crisp interval by α-cut operation, interval A_α shall be obtained as follows $\forall \alpha \in [0, 1]$

from

$$\frac{a_1^{(\alpha)} - a_1}{a_2 - a_1} = \alpha, \quad \frac{a_3 - a_3^{(\alpha)}}{a_3 - a_2} = \alpha$$

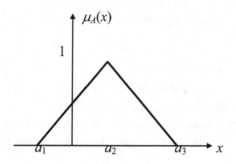

Fig. 5.6. Triangular fuzzy number $A = (a_1, a_2, a_3)$

we get

$$a_1^{(\alpha)} = (a_2 - a_1)\alpha + a_1$$
$$a_3^{(\alpha)} = -(a_3 - a_2)\alpha + a_3$$

thus

$$A_\alpha = [a_1^{(\alpha)}, a_3^{(\alpha)}]$$
$$= [(a_2 - a_1)\alpha + a_1, -(a_3 - a_2)\alpha + a_3]$$

Example 5.7 In the case of the triangular fuzzy number A = (−5, −1, 1) (Fig 5.7), the membership function value will be,

$$\mu_{(A)}(x) = \begin{cases} 0, & x < -5 \\[2mm] \dfrac{x+5}{4}, & -5 \leq x \leq -1 \\[2mm] \dfrac{1-x}{2}, & -1 \leq x \leq 1 \\[2mm] 0, & x > 1 \end{cases}$$

Fig. 5.7. $\alpha = 0.5$ cut of triangular fuzzy number $A = (-5, -1, 1)$

α-cut interval from this fuzzy number is

$$\frac{x+5}{4} = \alpha \implies x = 4\alpha - 5$$

$$\frac{1-x}{2} = \alpha \implies x = -2\alpha + 1$$

$$A_\alpha = [a_1^{(\alpha)}, a_3^{(\alpha)}] = [4\alpha - 5, -2\alpha + 1]$$

if $\alpha = 0.5$, substituting 0.5 for α, we get $A_{0.5}$

$$A_{0.5} = [a_1^{(0.5)}, a_3^{(0.5)}] = [-3, 0] \quad \square$$

5.3.2 Operation of Triangular Fuzzy Number

Same important properties of operations on triangular fuzzy number are summarized
(1) The results from addition or subtraction between triangular fuzzy numbers result also triangular fuzzy numbers.
(2) The results from multiplication or division are not triangular fuzzy numbers.
(3) Max or min operation does not give triangular fuzzy number.
but we often assume that the operational results of multiplication or division to be TFNs as approximation values.

1) Operation of triangular fuzzy number
first, consider addition and subtraction. Here we need not use membership function. Suppose triangular fuzzy numbers A and B are defined as,

$$A = (a_1, a_2, a_3), B = (b_1, b_2, b_3)$$

  i) Addition

$$
\begin{aligned}
A(+)B &= (a_1, a_2, a_3)(+)(b_1, b_2, b_3) \quad \text{: triangular fuzzy number} \\
&= (a_1 + b_1, a_2 + b_2, a_3 + b_3)
\end{aligned}
$$

ii) Subtraction

$$
\begin{aligned}
A(-)B &= (a_1, a_2, a_3)(-)(b_1, b_2, b_3) \quad \text{: triangular fuzzy number} \\
&= (a_1 - b_3, a_2 - b_2, a_3 - b_1)
\end{aligned}
$$

iii) Symmetric image

$$-(A) = (-a_3, -a_2, -a_1) \quad \text{: triangular fuzzy number}$$

Example 5.8 Let's consider operation of fuzzy number A, B(Fig 5.8).

$$A = (-3, 2, 4), B = (-1, 0, 6)$$
$$A (+) B = (-4, 2, 10)$$
$$A (-) B = (-9, 2, 5) \quad \square$$

2) Operations with α-cut

Example 5.9 α-level intervals from α-cut operation in the above two triangular fuzzy numbers A and B are

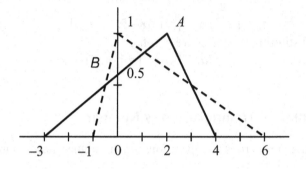

(a) Triangular fuzzy number A, B

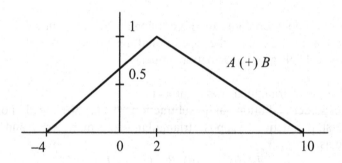

(b) $A (+) B$ of triangular fuzzy numbers

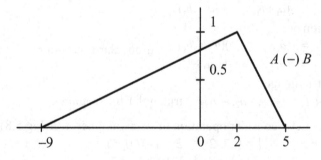

(c) $A (-) B$ triangular fuzzy numbers

Fig. 5.8. $A (+) B$ and $A (-) B$ of triangular fuzzy numbers

$$A_\alpha = [a_1^{(\alpha)}, a_3^{(\alpha)}] = [(a_2 - a_1)\alpha + a_1, -(a_3 - a_2)\alpha + a_3]$$
$$= [5\alpha - 3, -2\alpha + 4]$$
$$B_\alpha = [b_1^{(\alpha)}, b_3^{(\alpha)}] = [(b_2 - b_1)\alpha + b_1, -(b_3 - b_2)\alpha + b_3]$$
$$= [\alpha - 1, -6\alpha + 6]$$

performing the addition of two α-cut intervals A_α and B_α,

$$A_\alpha (+) B_\alpha = [6\alpha - 4, -8\alpha + 10]$$

especially for $\alpha = 0$ and $\alpha = 1$,

$$A_0 (+) B_0 = [-4, 10]$$
$$A_1 (+) B_1 = [2, 2] = 2$$

three points from this procedure coincide with the three points of triangular fuzzy number (-4, 2, 10) from the result $A(+)B$ given in the previous example.

Likewise, after obtaining $A_\alpha(-)B_\alpha$, let's think of the case when $\alpha = 0$ and $\alpha = 1$

$$A_\alpha (-) B_\alpha = [11\alpha - 9, -3\alpha + 5]$$

substituting $\alpha = 0$ and $\alpha = 1$ for this equation,

$$A_0 (-) B_0 = [-9, 5]$$
$$A_1 (-) B_1 = [2, 2] = 2$$

these also coincide with the three points of $A(-)B = (-9, 2, 5)$. □

Consequently, we know that we can perform operations between fuzzy number using α-cut interval.

5.3.3 Operation of General Fuzzy Numbers

Up to now, we have considered the simplified procedure of addition and subtraction using three points of triangular fuzzy number. However, fuzzy numbers may have general form, and thus we have to deal the operations with their membership functions.

Example 5.10 Addition $A (+) B$
Here we have two triangular fuzzy numbers and will calculate the addition operation using their membership functions

$$A = (-3, 2, 4), B = (-1, 0, 6)$$

$$\mu_{(A)}(x) = \begin{cases} 0, & x < -3 \\[2mm] \dfrac{x+3}{2+3}, & -3 \le x \le 2 \\[3mm] \dfrac{4-x}{4-2}, & 2 \le x \le 4 \\[2mm] 0, & x > 4 \end{cases}$$

$$\mu_{(B)}(y) = \begin{cases} 0, & y < -1 \\[2mm] \dfrac{y+1}{0+1}, & -1 \le y \le 0 \\[2mm] \dfrac{6-y}{6-0}, & 0 \le y \le 6 \\[2mm] 0, & y > 6 \end{cases}$$

for the two fuzzy number $x \in A$ and $y \in B$, $z \in A (+) B$ shall be obtained by their membership functions.

Let's think when $z = 8$. Addition to make $z = 8$ is possible for following cases :

$$2 + 6, \ 3 + 5, \ 3.5 + 4.5, \ \ldots$$

so

$$\begin{aligned} \mu_{A(+)B} &= \underset{8=x+y}{\vee} \ [\mu_A(2) \wedge \mu_B(6), \ \mu_A(3) \wedge \mu_B(5), \ \mu_A(3.5) \wedge \mu_B(4.5), \cdots] \\ &= \vee \ \ [1 \wedge 0, \ 0.5 \wedge 1/6, \ 0.25 \wedge 0.25, \cdots] \\ &= \vee \ \ [0, 1/6, \ 0.25, \cdots] \end{aligned}$$

If we go on these kinds of operations for all $z \in A (+) B$, we come to the following membership functions, and these are identical to the three point expression for triangular fuzzy number $A = (-4, 2, 10)$.

$$\mu_{A(+)B}(z) = \begin{cases} 0, & z < -4 \\[2mm] \dfrac{z+4}{6}, & -4 \le z \le 2 \\[2mm] \dfrac{10-z}{8}, & 2 \le z \le 10 \\[2mm] 0, & z > 10 \end{cases}$$ □

There in no simple method using there point expression for multiplication or division operation. So it is necessary to use membership functions.

Example 5.11 Multiplication $A (\bullet) B$
Let triangular fuzzy numbers A and B be

$$A = (1, 2, 4), \ B = (2, 4, 6)$$

$$\mu_{(A)}(x) = \begin{cases} 0, & x < 1 \\[2mm] x - 1, & 1 \le x < 2 \\[2mm] -\dfrac{1}{2}x + 2, & 2 \le x < 4 \\[2mm] 0, & x \ge 4 \end{cases}$$

$$\mu_{(B)}(y) = \begin{cases} 0, & y < 2 \\ \dfrac{1}{2}y - 1, & 2 \le y < 4 \\ -\dfrac{1}{2}y + 3, & 4 \le y < 6 \\ 0, & y \ge 6 \end{cases}$$

calculating multiplication $A\ (\bullet)\ B$ of A and B, $z = x \bullet y = 8$ is possible when $z = 2 \bullet 4$ or $z = 4 \bullet 2$

$$\begin{aligned} \mu_{A(\bullet)B} &= \underset{x\bullet y=8}{\vee}[\mu_A(2) \wedge \mu_B(4),\ \mu_A(4) \wedge \mu_B(2),\cdots] \\ &= \vee[1 \wedge 1,\ 0 \wedge 0,\cdots] \\ &= 1 \end{aligned}$$

also when $z = x \bullet y = 12$, $3 \bullet 4$, $4 \bullet 3$, $2.5 \bullet 4.8$, ... are possible.

$$\begin{aligned} \mu_{A(\bullet)B} &= \underset{x\bullet y=12}{\vee}[\mu_A(3) \wedge \mu_B(4),\ \mu_A(4) \wedge \mu_B(3),\ \mu_A(2.5) \wedge \mu_B(4.8),\cdots] \\ &= \vee[0.5 \wedge 1,\ 0 \wedge 0.5,\ 0.75 \wedge 0.6,\ \cdots] \\ &= \vee[0.5,\ 0,\ 0.6,\ \cdots] \\ &= 0.6 \end{aligned}$$

From this kind of method, if we come by membership function for all $z \in A\ (\bullet)\ B$, we see fuzzy number as in Fig 5.9. However, since this shape is in curve, it is not a triangular fuzzy number. For convenience, we can express it as a triangular fuzzy number by approximating $A\ (\bullet)\ B$

$$A(\bullet)B \cong (2,\ 8,\ 24)$$

we can wee that two end points and one peak point are used in this approximation. □

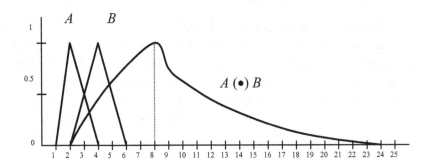

Fig. 5.9. Multiplication $A\ (\bullet)\ B$ of triangular fuzzy number

5.3.4 Approximation of Triangular Fuzzy Number

Since it is possible to express approximated values of multiplication and division as triangular fuzzy numbers, we are now up to the fact that how to get this approximated value easily.

Example 5.12 Approximation of multiplication
First, α-cuts of two fuzzy numbers are our main concern.

$$A = (1, 2, 4),\ B = (2, 4, 6)$$
$$\begin{aligned}
A_\alpha &= [(2-1)\alpha+1,\ -(4-2)\alpha+4] \\
&= [\alpha+1,\ -2\alpha+4] \\
B_\alpha &= [(4-2)\alpha+2,\ -(6-4)\alpha+6] \\
&= [2\alpha+2,\ -2\alpha+6]
\end{aligned}$$

For all $\alpha \in [0, 1]$, multiply A_α with B_α which are two crisp intervals. Now in $\alpha \in [0, 1]$, we see that elements of each interval are positive numbers. So multiplication operation of the two intervals is simple.

$$\begin{aligned}
A_\alpha(\bullet)B_\alpha &= [\alpha+1,\ -2\alpha+4](\bullet)[2\alpha+2,\ -2\alpha+6] \\
&= [(\alpha+1)(2\alpha+2),\ (-2\alpha+4)(-2\alpha+6)] \\
&= [2\alpha^2+4\alpha+2,\ 4\alpha^2-20\alpha+24]
\end{aligned}$$

when $\alpha = 0$,
$$A_0(\bullet)B_0 = [2,\ 24]$$

when $\alpha = 1$,
$$A_0(\bullet)B_1 = [2+4+2, 4-20+24] = [8, 8] = 8$$

we obtain a triangular fuzzy number which is an approximation of $A\ (\bullet)\ B$ (Fig 5.9).

$$A(\bullet)B \cong (2,\ 8,\ 24) \qquad \square$$

Example 5.13 Approximation of division
In the similar way, let's express approximated value of $A\ (/)\ B$ in a triangular fuzzy number. First, divide interval A_α by B_α. We reconsider the sets A and B in the previous example. For $\alpha \in [0, 1]$, since element in each interval has positive number, we get $A_\alpha\ (/)\ B_\alpha$ as follows.

$$A_\alpha(/)B_\alpha = [(\alpha+1)/(-2\alpha+6),\ (-2\alpha+6)/(2\alpha+2)]$$

when $\alpha = 0$,
$$\begin{aligned}
A_0(/)B_0 &= [1/6,\ 4/2] \\
&= [0.17,\ 2]
\end{aligned}$$

when $\alpha = 1$,
$$\begin{aligned}
A_1(/)B_1 &= [(1+1)/(-2+6),\ (-2+4)/(2+2)] \\
&= [2/4,\ 2/4] \\
&= 0.5
\end{aligned}$$

so the approximated value of $A\ (/)\ B$ will be

$$A(/)B = (0.17,\ 0.5,\ 2) \qquad \square$$

5.4 Other Types of Fuzzy Number

5.4.1 Trapezoidal Fuzzy Number

Another shape of fuzzy number is trapezoidal fuzzy number. This shape is originated from the fact that there are several points whose membership degree is maximum ($\alpha = 1$).

Definition (Trapezoidal fuzzy number) We can define trapezoidal fuzzy number A as

$$A = (a_1, a_2, a_3, a_4)$$

the membership function of this fuzzy number will be interpreted as follows(Fig 5.10).

$$\mu_A(x) = \begin{cases} 0, & x < a_1 \\ \dfrac{x - a_1}{a_2 - a_1}, & a_1 \leq x \leq a_2 \\ 1, & a_2 \leq x \leq a_3 \\ \dfrac{a_4 - x}{a_4}, & a_3 \leq x \leq a_4 \\ 0, & x > a_4 \end{cases}$$

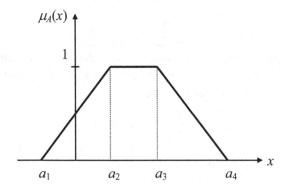

Fig. 5.10. Trapezoidal fuzzy number $A = (a_1, a_2, a_3, a_4)$

α-cut interval for this shape is written below.
$\forall \alpha \in [0, 1]$

$$A_\alpha = [(a_2 - a_1)\alpha + a_1, \quad -(a_4 - a_3)\alpha + a_4]$$

when $a_2 = a_3$, the trapezoidal fuzzy number coincides with triangular one.

5.4.2 Operations of Trapezoidal Fuzzy Number

Let's talk about the operations of trapezoidal fuzzy number as in the triangular fuzzy number,
(1)   Addition and subtraction between fuzzy numbers become trapezoidal fuzzy number.
(2)   Multiplication, division, and inverse need not be trapezoidal fuzzy number.
(3)   Max and Min of fuzzy number is not always in the form of trapezoidal fuzzy number.
    But in many cases, the operation results from multiplication or division are approximated trapezoidal shape. As in triangular fuzzy number, addition and subtraction are simply defined, and multiplication and division operations should be done by using membership functions.
(1) Addition

$$A(+)B \;=\; (a_1, a_2, a_3, a_4)(+)(b_1, b_2, b_3, b_4)$$
$$=\; (a_1 + b_1, a_2 + b_2, a_3 + b_3, a_4 + b_4)$$

(2) Subtraction

$$A(-)B \;=\; (a_1 - b_4, a_2 - b_3, a_3 - b_2, a_4 - b_1)$$

Example 5.14   Multiplication
Multiply two trapezoidal fuzzy numbers as following:

$$A = (1, 5, 6, 9)$$
$$B = (2, 3, 5, 8)$$

    For exact value of the calculation, the membership functions shall be used and the result is described in (Fig. 5.11) For the approximation of operation results, we use α-cut interval

$$A_\alpha = [4\alpha + 1, -3\alpha + 9]$$
$$B_\alpha = [\alpha + 2, -3\alpha + 8]$$

since, for all $\alpha \in [0, 1]$, each element for each interval is positive, multiplication between α-cut intervals will be

$$A_\alpha(\bullet)B_\alpha \;=\; [(4\alpha+1)(\alpha+2), \;\; (-3\alpha+9)(-3\alpha+8)]$$
$$=\; [4\alpha^2 + 9\alpha + 2, \;\; 9\alpha^2 - 51\alpha + 72]$$

if $\alpha = 0$,

$$A_0(\bullet)B_0 \;=\; [2, \;\; 72]$$

if $\alpha = 1$,

$$A_1(\bullet)B_1 \;=\; [4 + 9 + 2, \;\; 9 - 51 + 72]$$
$$=\; [15, \;\; 30]$$

so using four points in $\alpha = 0$ and $\alpha = 1$, we can visualize the approximated value as trapezoidal fuzzy number as (Fig. 5.11)

$$A(\bullet)B \cong [2, \quad 15, \quad 30, \quad 72] \quad \square$$

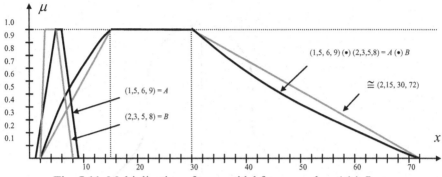

Fig. 5.11. Multiplication of trapezoidal fuzzy number A ($\bullet$) B

Generalizing trapezoidal fuzzy number, we can get flat fuzzy number. In other words, flat fuzzy number is for fuzzy number A satisfying following

$$\exists m_1, m_2 \in \Re, \qquad m_1 < m_2$$
$$\mu_A(x) = 1, \qquad m_1 \leq x \leq m_2$$

In this case, not like trapezoidal form, membership function in $x < m_1$ and $x < m_2$ need not be a line as shown in (Fig 5.12.)

5.4.3 Bell Shape Fuzzy Number

Bell shape fuzzy number is often used in practical applications and its function is defined as follows(Fig 5.13)

$$\mu_f(x) = \exp\left\{\frac{-(x-m_f)^2}{2\delta_f^2}\right\}$$

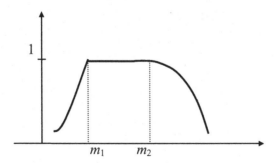

Fig. 5.12. Flat fuzzy number

where μ_f is the mean of the function, δ_f is the standard deviation.

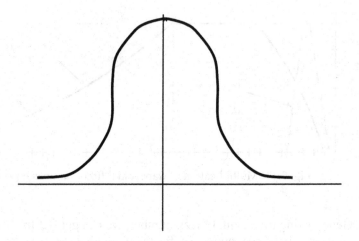

Fig. 5.13. Bell shape fuzzy number

[SUMMARY]

☐ Interval and Fuzzy number
  – Interval is a crisp set of continuous numbers
  – Fuzzy number is an interval whose boundary is ambiguous

☐ Operations of interval and fuzzy number
  – Operation of interval
  – α-cut of fuzzy number becomes an interval

☐ Operation of fuzzy number
  – By using α-cut
  – By using membership function

☐ Triangular fuzzy number
  – Three point expression (a_1, a_2, a_3)
  – α-cut of triangular fuzzy number $[(a_2 - a_1)\,\alpha + a_1, (a_3 - a_2)\,\alpha + a_3]$

☐ Operation of triangular fuzzy number
  – Addition and subtraction between triangular fuzzy numbers result triangular numbers.
  – Multiplication and division do not give triangular number.
  – Approximation of multiplication

☐ Trapezoidal fuzzy number
  – Four point expression (a_1, a_2, a_3, a_4)
  – α-cut of trapezoidal fuzzy number
  – Operation of trapezoidal number
  – Approximation of multiplication

☐ Bell shape fuzzy number
  – Bell shape membership function
  –
$$\mu_f(x) = \exp\left\{\frac{-(x - m_f)^2}{2\delta_f^2}\right\}$$
  – m_f : mean of the function
  – δ_f : standard deviation

[EXERCISES]

5.1 Calculate the following operations of intervals
   a)   [5, 8] (+) [-1, 3]
   b)   [5, 9] (-) [5, 10]
   c)   [6, 10] (•)[-1, 5]
   d)   [3, 7] (/) [-3, -1]
   e) $[5, 8]^{-1}$

5.2 Calculate the operations of fuzzy sets A={(5, 0.5), (6, 1.0)}, B ={(2, 1.0), (3, 0.4), (4, 0.8)}
   a)   A (+) B
   b)   A (-) B
   c)   A (•) B
   d)   A (∨) B
   e)   A(/) B

5.3 Calculate α-cut interval of triangular fuzzy number
   A = (-4, 2, 5)

5.4 Calculate the following operations of triangular fuzzy sets by using α-cut operation
           A = (1, 3, 8), B (2, 4, 5)
   a)   A (+) B
   b)   A (-) B
   c)   A (•) B
   d)   A(/) B
   e) A^{-1}
   f)   A (∨) B
   g)   A (∧) B

5.5 Show α-cut of trapezoidal fuzzy number A = (-4, -1, 2, 5)

5.6 Calculate the operations of trapezoidal fuzzy number by using α-cut operation
   a)   A (+) B
   b)   A (-) B
   c)   A (•) B
   d)   A(/) B

e) A^{-1}

f) $A\,(\vee)\,B$

g) $A\,(\wedge)\,B$

Chapter 6. FUZZY FUNCTION

We introduce the concept of fuzzy function. Fuzzy functions consist of crisp function with fuzzy constraint and fuzzifying function. Maximizing set and minimizing set are also introduced and applied to find the maximum value with fuzzy domain of crisp function. In the later part, fuzzy integration and differentiation are discussed with examples.

6.1 Kinds of Fuzzy Function

Fuzzy function can be classified into following three groups according to which aspect of the crisp function the fuzzy concept was applied.
(1) Crisp function with fuzzy constraint.
(2) Crisp function which propagates the fuzziness of independent variable to dependent variable.
(3) Function that is itself fuzzy. This fuzzifying function blurs the image of a crisp independent variable.

6.1.1 Function with Fuzzy Constraint

Definition (Function with fuzzy constraint) Let X and Y be crisp sets, and f be a crisp function. A and B are fuzzy sets defined on universal sets X and Y respectively. Then the function satisfying the condition $\mu_A(x) \leq \mu_B(f(x))$ is called a function with constraints on fuzzy domain A and fuzzy range B.

$$f : X \to Y. \qquad \square$$

Example 6.1 There is a function $\quad y = f(x)$
Assume that the function f has fuzzy constraint like this,
$\qquad$ "If x is a member of A, then y is a member in B"
"The membership degree $\mu_A(x)$ of x for A is less than that $\mu_B(y)$ of y for B"
or
$$"\mu_A(x) \leq \mu_B(y)".$$
The previous fuzzy constraints denote the sufficient fuzzy condition for y to be a member of B.

"If membership degree of x for A is α, then that of y for B would be no less than α"   □

Example 6.2 Consider two fuzzy sets,
$$A = \{(1, 0.5), (2, 0.8)\}, \quad B = \{(2, 0.7), (4, 0.9)\}$$
and a function
$$y = f(x) = 2x, \quad \text{for} \quad x \in A, \quad y \in B.$$
We see the function f satisfies the condition, $\mu_A(x) \leq \mu_B(y)$.   □

Example 6.3 We shall investigate a function with the following statement.
          " A competent salesman gets higher income"
Let X and Y be sets of salesmen and of monthly income $[0, \infty]$ respectively. And A and B are fuzzy sets of "competent salesmen" and "high income". In this case, the functions f
$$f : A \rightarrow B$$
satisfies the following for all $x \in A$ and $y = f(x) \in B$
$$\mu_A(x) \leq \mu_B(f(x)).$$ □
Consider a function satisfying fuzzy constraint $f : A \rightarrow B$, $g : B \rightarrow C$ (A, B and C denote fuzzy sets defined on X, Y and Z). The composition of these two functions yields fuzzy function with fuzzy constraint.
$$g \circ f : A \rightarrow C.$$
That is due to the conditions $\mu_A(x) \leq \mu_B(f(x))$, $\mu_B(y) \leq \mu_C(g(y))$ and $y = f(x)$, $z = g(y)$. The following holds.
$$\mu_A(x) \leq \mu_C(g(f(x))).$$

6.1.2 Propagation of Fuzziness by Crisp Function

Definition (Fuzzy extension function) Fuzzy extension function propagatetes the ambiguity of independent variables to dependent variables. when f is a crisp function from X to Y, the fuzzy extension function f defines the image $f(\tilde{X})$ of fuzzy set $\tilde{X}$. That is, the extension principle is applied(see section 3.4).
$$\mu_{f(\tilde{x})}(y) = \begin{cases} \max\limits_{x \in f^{-1}(y)} \mu_{\tilde{x}}(x), & if \quad f^{-1}(y) \neq \phi \\ 0, & if \quad f^{-1}(y) = \phi \end{cases}$$
where, $f^{-1}(y)$ is inverse image of y.
In this section, we use the sign $\sim$ for the emphasis of fuzzy variable.

Example 6.4 There is a crisp function,
$$f(x) = 3\tilde{x} + 1$$
where its domain is $A = \{(0, 0.9), (1, 0.8), (2, 0.7), (3, 0.6), (4, 0.5)\}$ and its range is $B = [0, 20]$
The independent variables have ambiguity and the fuzziness is propagated to the crisp set B. Then, we can obtain a fuzzy set B' in B

$$B' = \{(1,0.9), (4, 0.8), (7, 0.7), (10, 0.6), (13, 0.5)\}. \quad \square$$

There are another examples of the fuzzy extension function.

$$\tilde{y} = a\tilde{x} + b\tilde{x}^2$$
$$\tilde{y} = a\cos\tilde{x} + b$$

6.1.3 Fuzzifying Function of Crisp Variable

Fuzzifying function of crisp variable is a function which produces image of crisp domain in a fuzzy set.

Definition (Single fuzzifying fuction) Fuzzifying function from X to Y is the mapping of X in fuzzy power set $\tilde{P}(Y)$.

$$\tilde{f} : X \rightarrow \tilde{P}(Y) \quad \square$$

That is to say, the fuzzifying function is a mapping from domain to fuzzy set of range. Fuzzifying function and the fuzzy relation coincides with each other in the mathematical manner. So to speak, fuzzifying function can be interpreted as fuzzy relation R defined as following:

$$\forall (x, y) \in X \times Y$$
$$\mu_{\tilde{f}(x)}(y) = \mu_R(x, y)$$

Example 6.5 Consider two crisp sets $A = \{2, 3, 4\}$ and $B = \{2, 3, 4, 6, 8, 9, 12\}$

A fuzzifying function $\tilde{f}$ maps the elemets in A to power set $\tilde{P}(B)$ in the following manner.

$$\tilde{f}(2) = B_1, \quad \tilde{f}(3) = B_2, \quad \tilde{f}(4) = B_3$$

$$\text{where} \quad \tilde{P}(B) = \{B_1, B_2, B_3\}$$

$B_1 = \{(2, 0.5), (4, 1), (6. 0.5)\} \quad B_2 = \{(3. 0.5), (6,1), (9, 0.5), \quad B_3 = \{(4, 0.5), (8, 1), (12, 0.5)\}$

If we look at the mapping in detail, we can see the relationship as shown in (Fig 6.1)

The function $\tilde{f}$ maps element $2 \in A$ to element $2 \in B_1$ with degree 0.5, to element $4 \in B_1$ with 0.1, and to element $6 \in B_1$ with 0.5. Now we apply a-cut operation to the fuzzifying function.

$$f : 2 \rightarrow \{2, 4, 6\} \quad \text{for} \quad \alpha = 0.5$$
$$f : 2 \rightarrow \{4\} \quad \text{for} \quad \alpha = 1.0$$

In the same manner

$$+f : 3 \rightarrow \{3, 6, 9\} \quad \text{for} \quad \alpha = 0.5$$
$$f : 3 \rightarrow \{6\} \quad \text{for} \quad \alpha = 1.0$$

again

$$f : 4 \rightarrow \{4, 8, 12\} \quad \text{for} \quad \alpha = 0.5$$
$$f : 4 \rightarrow \{8\} \quad \text{for} \quad \alpha = 1.0 \quad \square$$

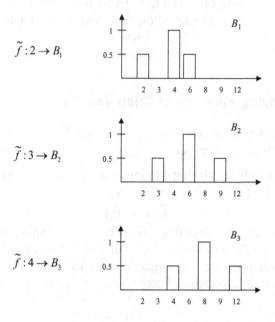

$\tilde{f}:2 \to B_1$

$\tilde{f}:3 \to B_2$

$\tilde{f}:4 \to B_3$

Fig. 6.1. Fuzzifying function

Definition (Fuzzy bunch of functions) Fuzzy bunch of crisp functions from X to Y is defined with fuzzy set of crisp function f_i ($i = 1, \ldots, n$) and it is denoted as

$$\tilde{f} = \{(f_i, \mu_{\tilde{f}}(f_i)) \mid f_i : X \to Y, i \in \aleph\}$$

$$f_i = f(x), \quad \forall x \in X \qquad \square$$

This function produces fuzzy set as its outcome.

Example 6.6 In the case of crisp sets f_1, f_2 and f_3, the bunch will be, for example,

$X = \{1, 2, 3\}$

$$\tilde{f} = \{(f_1, 0.4), (f_2, 0.7), (f_3, 0.5)\}$$

$$f_1(x) = x, \quad f_2(x) = x^2, \quad f_3(x) = -x +1$$

By f_1, we get $\tilde{f}_1 = \{(1, 0.4), (2, 0.4), (3, 0.4)\}$

By f_2, $\tilde{f}_2 = \{(1, 0.7), (4, 0.7), (9, 0.7)\}$

By f_3, $\tilde{f}_3 = \{(0, 0.5), (-1, 0.5), (-2, 0.5)\}$

then, we can summarize the outputs as follows :

$\tilde{f}(1) = \{(1, 0.4), (1, 0.7), (0, 0.5)\} = \{(0, 0.5), (1, 0.7)\}$

$\tilde{f}(2) = \{(2, 0.4), (4, 0.7), (-1, 0.5)\} = \{(-1, 0.5), (2, 0.4), (4, 0.7)\}$

$\tilde{f}(3) = \{(3, 0.4), (9, 0.7), (-2, 0.5)\} = \{(-2, 0.5), (3, 0.4), (9, 0.7)\}$

We can see that the fuzzy function maps 2 to 2 with possibility 0.4 through f_1, to 4 with 0.7 through f_2 and to -1 with 0.5 through f_3. This result is represented by the above $\tilde{f}_2(2)$.   □

Example 6.7 There is a fuzzy bunch of continuous functions (Fig 6.2).

$$\tilde{f} = \{(f_1, 0.4), (f_2, 0.7), (f_3, 0.5)\}$$

$X = [1, 2]$

$$f_1(x) = x, \quad f_2(x) = x^2, \quad f_3(x) = x^2+1$$

This fuzzy function maps 1.5 to 1.5 with possibility 0.4 through f_1, to 2.25 with 0.7 through f_2, and to 3.25 with 0.5 through f_3.

that is $\tilde{f}(1.5) = \{(1.5, 0.4), (2.25, 0.7), (3.25, 0.5)\}$ □

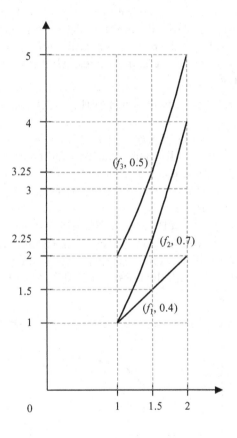

Fig. 6.2. Fuzzy bunch of function

6.2 Fuzzy Extrema of Function

6.2.1 Maximizing and Minimizing Set

Definition (Maximizing set) Let f be the function having real values in X and the highest and the lowest value of f be $sup(f)$ and $inf(f)$ respectively. At this time, the maximizing set M is defined as a fuzzy set.

$$\forall x \in X, \qquad \mu_M(x) = \frac{f(x) - \inf(f)}{\sup(f) - \inf(f)} \qquad \square$$

That is, the maximizing set M is a fuzzy set and defined by the possibility of x to make the maximum value $sup(f)$. The possibility of x to be in the range of M is defined from the relative normalized position in the interval $[\inf(f), \sup(f)]$. Here the interval $[\inf(f), \sup(f)]$ denotes the possible range of $f(x)$ to have some values. Minimizing set of f is defined as the maximizing set of $-f$.

Example 6.8 Let's have a look at $f(x)$ of (Fig 6.3.) The interval of values is as follows :

$$[\inf(f), \sup(f)] = [10, 20], \ 1 \le x \le 10$$

and when $x = 5$, $f(x) = 15$. Then the possibility of $x = 5$ to be in the maximizing set M is calculated as follows :

$$\mu_M(5) = (15 - 10) / (20 - 10) = 5 / 10 = 0.5$$

also if when $x = 8$, $f(x) = 19$,

$$\mu_M(8) = (19 - 10) / (20 - 10) = 9 / 10 = 0.9$$

$\mu_M(x)$ denotes the possibility of x to make maximum value of f. Here, we might say that two independent variables $x = 5$ and $x = 8$ make the maximum value of $f(x) = 20$ with the possibilities 0.5 and 0.9, respectively $\square$

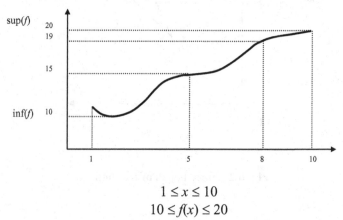

$$1 \le x \le 10$$
$$10 \le f(x) \le 20$$

Fig. 6.3. Example of maximizing set

Example 6.9 Following example is to obtain the maximizing set M of $f(x)$ = $\sin x$ ($0 \leq x \leq 2\pi$) (Fig. 6.4).

$$\mu_M(x) = \frac{\sin x - \inf(\sin x)}{\sup(\sin x) - \inf(\sin x)}$$

$$= \frac{\sin x - (-1)}{1 - (-1)}$$

$$= \frac{\sin x + 1}{2}$$

$$= \frac{1}{2}\sin x + \frac{1}{2}$$

If $x = \pi$ for example, $f(x) = \sin \pi = 0$. The possibility for $f(x) = 0$ to be the maximum value of sin function is $1/2$.   □

6.2.2 Maximum Value of Crisp Function

(1)   Crisp Domain
Assume x_0 is the independent variable which makes function f be the maximum value in crisp domain D. We might utilize the maximizing set M to find the value x_0. That is, x_0 shall be the element that enables $\mu_M(x)$ to be the maximum value.

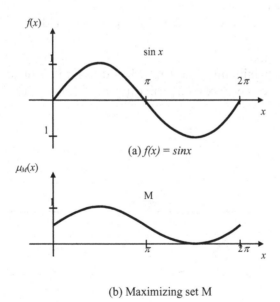

(a) $f(x) = sinx$

(b) Maximizing set M

Fig. 6.4. Maximizing set of sin function

$$\mu_M(x_0) \;=\; \max_{x \in D}\mu_M(x)$$

$\mu_M(x)$ is the membership function of maximizing set. At this time, maximum value of f will be $f(x_0)$. $\mu_M(x_0)$ can be written as in the following, denoting domain D as a crisp set.

$$\begin{aligned}\mu_M(x_0) \;&=\; \max_{x \in D}\mu_M(x)\\ &=\; \max_{x \in X}\min[\mu_M(x),\mu_D(x)]\end{aligned}$$

Note that domain D is replaced by universal set X in the above formula. We expressed the possibility of x to be in D as $\mu_D(x)$.

Example 6.10 There is a function (F.g. 6.5) and its domain.

$$f(x) = \cos x, \quad x \in D = [0,\ 2\pi]$$

$$\mu_M(x) = \frac{\cos x - \inf(\cos x)}{\sup(\cos x) - \inf(\cos x)} = \frac{\cos x - (-1)}{1-(-1)} = \frac{1}{2}\cos x + \frac{1}{2}$$

$$\mu_D(x) = 1 \quad \text{for } 0 \le x \le 2\pi,$$
$$= 0 \quad \text{otherwise}$$

Maximum value $f(x_0)$ is obtained at x_0
where

$$\begin{aligned}\mu_M(x_0) \;&=\; Max\,Min[\mu_M(x),\mu_D(x)]\\ &=\; \underset{0 \le x \le 2\pi}{Max}\ \mu_M(x),\\ &=\; 1 \quad \text{when}\quad x_0 = 0 \ \text{ and } 2\pi\end{aligned}$$

Therefore, the maximum value
$$f(x_0) \;=\; 1 \quad \text{is obtained when}\quad x_0 = 0 \ \text{ and } 2\pi \qquad \square$$

(2) Fuzzy Domain
Now getting the maximum value $f(x_0)$ when domain is expressed in fuzzy set. To make f be the maximum value by x_0, following two conditions should be met.

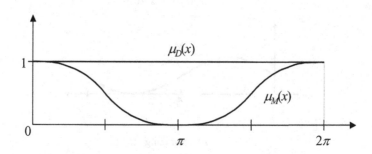

Fig. 6.5. Maximum value with crisp domain

- Set $\mu_M(x)$ as maximum
- Set $\mu_D(x)$ as maximum

For arbitrary element x_1 corresponding to the maximum f, it is necessary to satisfy the above two conditions on $\mu_M(x)$ and $\mu_D(x)$. The possibility of x_1 to make the maxim value of f is determined by the minimum of $\mu_M(x_1)$ and $\mu_D(x_1)$, that is,

$$\text{Min}[\mu_M(x_1), \mu_D(x_1)].$$

Therefore, the point x_0 which enables the function f to be the maximum is defined as follows.

$$\underset{x \in X}{MaxMin}[\mu_M(x), \mu_D(x)] = \mu(x_0)$$

At this time, the maximum value is $f(x_0)$. Here $\mu_M(x)$ is membership function of maximizing set and $\mu_D(x)$ is that of fuzzy domain(Fig 6.6). Comparing x_0 with x_1 in the figure, x_1 enables f to be maximum rather than x_0.

$$f(x_1) > f(x_0) \text{ or } \mu_M(x_1) > \mu_M(x_0)$$

but since $\mu_D(x_1)$ is very much smaller than $\mu_D(x_0)$, $f(x_0)$ is selected as the maximum value.

Example 6.11 There are a function and a fuzzy domain(Fig 6.7).

$$f(x) = -x + 2, \ x \in D$$
$$\mu_D(x) = x^2 \ \text{ for } \ 0 \le x \le 1$$
$$= 0 \quad \text{otherwise}$$

We can get the maximizing function.

$$\mu_M(x) = \frac{-x + 2 - 1}{2 - 1} = -x + 1$$

From the following equation,

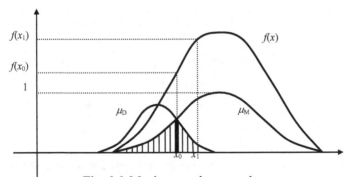

$$f(x_1) \cdots \qquad \qquad f(x)$$
$$f(x_0) \cdots$$
$$1 \cdots$$
$$\mu_D \qquad \qquad \mu_M$$

Fig. 6.6. Maximum value as scalar

$$\mu_f(x_0) = \text{Max Min}[\mu_M(x), \ \mu_D(x)]$$

the point x_0 is obtained when

$$\mu_M(x) = \mu_D(x) \quad \text{for} \ \ 0 \le x \le 1$$

$$-x+1 = x^2, \quad x \cong 0.6$$

Therefore, we have the maximum value $f(x_0) = 1.4$ when $x_0 = 0.6$ ◻

Example 6.12 We have a crisp function f and its fuzzy domain D. Let's find the maximum value of f with D.

$$f(x) = \cos x, \quad x \in D$$

$$\mu_D(x) = \text{Min} \ [1, \frac{x}{\pi}] \quad \text{for} \ \ 0 \le x \le 2\pi$$

$$= 0 \qquad\qquad\qquad \text{otherwise}$$

$$\mu_M(x) = \frac{1}{2}\cos x + \frac{1}{2}$$

In (Fig 6.8.)

Max Min $[\mu_M(x), \mu_D(x)]$ is obtained when $x_0 = 2\pi$

then $f(x_0) = 1$

$$\mu_M(x_0) = \mu_D(x_0) = 1 \qquad ◻$$

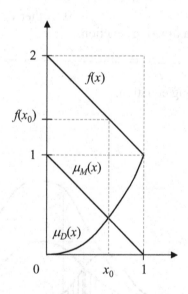

Fig. 6.7. Maximum value of example $f(x) = -x+2$ with fuzzy domain

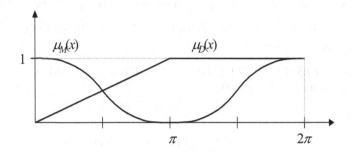

Fig. 6.8. Maximum value $f(x) = \cos(x)$ with fuzzy domain

6.3 Integration and Differenciation of Fuzzy Function

6.3.1 Integration

In this section, we are up to the task of integration of fuzzifying function in non-fuzzy interval and that of such function in fuzzy interval.

(1) Integration of fuzzifying function in crisp interval

Definition (Integration of fuzzying function) In non-fuzzy interval $[a, b]$ $\in \Re$, let the fuzzifying function have fuzzy value $\tilde{f}(x)$ for $x \in [a, b]$. Integration $\tilde{I}(a,b)$ of the fuzzifying function in $[a, b]$ is defined as follows:

$$\tilde{I}(a,b) = \{(\int_a^b f_\alpha^-(x)dx + \int_a^b f_\alpha^+(x)dx, \ \alpha) \mid \alpha \in [0,1]\}$$

Here f_α^+ and f_α^- are α-cut functions of $\tilde{f}(x)$. Note that the plus sign(+) in the above formula is to express enumeration in fuzzy set but not addition. Therefore, the total integration is obtained by aggregating integrations of each α-cut function.

If we apply the a-cut operation to the fuzzifying function, we can get f_α^+ or f_α^- which are α-cut functions. We can calculate the integration of each function :

$$\tilde{I}_a^- = \int_a^b f_\alpha^-(x)dx \quad \text{and} \quad \tilde{I}_a^- = \int_a^b f_\alpha^+(x)dx.$$

Now we can say that the possibility of $\tilde{I}_a^-$ or $\tilde{I}_a^+$ to be a member of total integration $\tilde{I}(a,b)$ is α. Recall the principle in calculating the fuzzy cardinality in sec 1.5.5.

Example 6.13 There is a fuzzy bunch of functions and we want to get integration in [1, 2](Fig 6.9).

$$\tilde{f} = \{(f_1, 0.4), (f_2, 0.7), (f_3, 0.4)\}$$

$X = [1, 2]$

$$f_1(x) = x, \quad f_2(x) = x^2, \quad f_3(x) = x+1$$

i) Integration at $\alpha = 0.7$,
$$f = f_2(x) = x^2$$

$$I_a(1,2) = \int_1^2 x^2 dx = \frac{1}{3} x^3 \Big]_1^2 = \frac{7}{3}$$

The integration result is $\frac{7}{3}$ with possibility 0.7

Therefore , $\tilde{I}_{0.7}(1,2) = \{(\frac{7}{3}, 0.7)\}$

ii) $\alpha = 0.4$, there are two functions
$$f^+ = f_1(x) = x$$
$$f^- = f_3(x) = x+1$$

$$I_a^+(1, 2) = \int_1^2 x \, dx = \frac{1}{2} x^2 \Big]_1^2 = \frac{3}{2}$$

$$I_a^-(1, 2) = \int_1^2 (x+1) dx = \frac{1}{2} x^2 + x \Big]_1^2 = \frac{5}{2}$$

The integration results are $\frac{3}{2}$ with possibility 0.4 and $\frac{5}{2}$ with 0.4.

then, $\tilde{I}_{0.4}(1, 2) = \{(\frac{3}{2}, 0.4), (\frac{5}{2}, 0.4)\}$

Finally, we have the total integration.

$$\tilde{I}(1, 2) = \{(\frac{7}{3}, 0.7), (\frac{3}{2}, 0.4), (\frac{5}{2}, 0.4)\} \qquad \square$$

(2) Integration crisp function in fuzzy interval
In this part, we shall deal with the integration of non-fuzzy function in fuzzy interval [A, B] of which the boundaries are determined by two fuzzy sets A and B.(Fig 6.10)

Definition (Integration in fuzzy interval) Integration $I(A, B)$ of non-fuzzy function f in fuzzy interval [A, B] is defined as,

$$\mu_{I(a,b)}(z) = \underset{\substack{x,y \\ z = \int_x^y f(u)du}}{Max} \; Min[\mu_A(x), \mu_B(x)]$$

$\square$

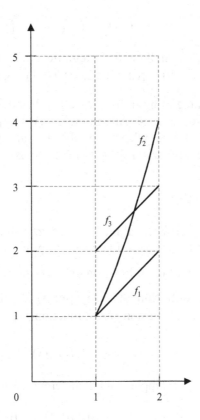

Fig. 6.9. Integration of fuzzy function with crisp interval

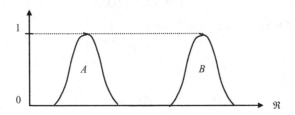

Fig. 6.10. Fuzzy interval

Example 6.14 Following shows the integration of function $f(x) = 2$ in fuzzy interval $[A, B]$.

$$A = \{(4, .8), (5, 1), (6, .4)\}$$
$$B = \{(6, .7), (7, 1), (8\ .2)\}$$
$$f(x) = 2, x \in [4, 8]$$

$$\tilde{I}(A, B) = \int_A^B f(x)dx = \int_A^B 2dx$$

Like (Table 7.1.), we shall get the integration value $I(A, B)$.
$$\tilde{I}(A, B) = \{(0, .4), (2, .7), (4, 1), (6, .8), (8, .2)\}$$

For instance, integrating in $[6, 6]$, we get 0 as the integration value. The possibility of this case is 0.4. And in the interval of $[5, 6]$ and $[6, 7]$, we get the integration value 2 whose possibilities are 0.7 and 0.4 . So the possibility for the integration value to be 2 is $\max[0.7, 0.4] = 0.7$.   □

6.3.2 Differentiation

Here we meet the differentiation of non-fuzzy function in fuzzy interval and that of fuzzifying function in non-fuzzy points.

(1)   Differentiation of crisp function on fuzzy points

Definition (Differentiation at fuzzy point) By the extension principle, differentiation $f'(A)$ of non-fuzzy function f at fuzzy point or fuzzy set A is defined as

$$\mu_{f'(A)}(y) = \underset{f(x)=y}{Max}\, \mu_A(x)   □$$

Example 6.15 For example, when differentiating function $f(x) = x^3$ at fuzzy point A,
$$A = \{(-1, 0.4), (0, 1), (1, 0.6)\}$$
from $f'(x) = 3x^2$,
$$f'(A) = \{(3, 0.4), (0, 1), (3, 0.6)\}$$
$$= \{(0, 1), (3, 0.6)\}   □$$

Example 6.16 There is a fuzzifying fuction
$$\tilde{f} = \{(f_1, 0.4), (f_2, 0.7), (f_3, 0.4)\}$$
$$f_1(x) = x, \quad f_2(x) = x^2, \quad f_3(x) = x^3+1$$
First, we have
$$f_1'(x) = 1 \quad f_2'(x) = 2x, \quad f_3'(x) = 3x^2$$
$$f_1'(0.5) = 1 \qquad \text{when} \quad \alpha = 0.4$$
$$f_2'(0.5) = 1 \qquad \text{when} \quad \alpha = 0.7$$
$$f_3'(0.5) = 0.75 \qquad \text{when} \quad \alpha = 0.4$$
$$\frac{d\tilde{f}}{dx}(x_0) = \{(1, 0.4), (1, 0.7), (0.75, 0.4)\}$$
$$= \{(1, 0.7), (0.75, 0.4)\}   □$$

Table 6.1. Fuzzy Integration

[a, b]	$\int_a^b 2\,dx$	$\min[\mu_A(a), \mu_B(b)]$
[4, 6]	4	.7
[4, 7]	6	.8
[4, 8]	8	.2
[5, 6]	2	.7
[5, 7]	4	1.0
[5, 8]	6	.2
[6, 6]	0	.4
[6, 7]	2	.4
[6, 8]	4	.2

[SUMMARY]

☐ Kinds of fuzzy function
  - Crisp function with fuzzy constraint
  - Crisp function reflects fuzziness of independent variable to dependent one
  - Fuzzifying function

☐ Fuzzifying function
  - Function giving ambiguous image of crisp independent variable
  - a-cut operation of fuzzifying function

☐ Fuzzy extreme of function
  - Maximizing fuzzy set
  - Minimizing fuzzy set

☐ Maximum value of crisp function
  - Crisp domain
  - Fuzzy domain

☐ Integration
  - Integration of fuzzifyng function in non-fuzzy interval
  - Integration of non-fuzzy function in fuzzy interval

☐ Differentiation
  - Differentiation of crisp function at fuzzy points
  - Differentiation of fuzzifying function at non-fuzzy points

[EXERCISES]

6.1 Show the following function satisfies the conditions.

Condition : $\mu_A(x) \le \mu_B(y)$

Function : $y = f(x) = 3x^2, \quad x \in A, \quad y \in B$

$$A = \{ (2, 0.5), (3, 0.4)\}$$
$$B = \{(4, 0.4), (12, 0.5), (27, 0.5)\}$$

6.2 Show the following function is a fuzzifying function.

$\tilde{f} : A \rightarrow \tilde{p}(B)$

$$\tilde{f}(1) = B_1$$
$$\tilde{f}(2) = B_2$$
$$\tilde{f}(3) = B_3$$

where $A = \{1, 2, 3\} \quad B = \{1, 2, 3, 4, 6\}$

$\tilde{p}(B) = (B_1, B_2, B_3)$

$$B_1 = \{(1, 0.9), (2, 0.5)\}$$
$$B_2 = \{(2, 0.5), (4, 0.9)\}$$
$$B_3 = \{(3, 1.0), (6, 0.5)\}$$

6.3 There in a fuzzy bunch of function

$X = \{2, 3, 4\}$

$$f_1(x) = x+1, \quad f_2(x) = x^2, \quad f_3(x) = x^2+1$$
$$\tilde{f} = \{(f_1, 0.4), (f_2, 0.5), (f_3, 0.9)\}$$

Finds $\tilde{f}(2)$, $\tilde{f}(3)$ and $\tilde{f}(4)$.

6.4 There is a function $f_3(x) = x^3, \quad x \in D = [-1, 3]$

Define its maximizing fuzzy set and minimizing fuzzy set.
Calculate the possibilities of $x = 0$ in each set.

6.5 Find maximizing set and minimizing set of $f(x) = \cos x$

Calculate the possibilities of $\dfrac{\pi}{2}$ and 2π making the maximum and minimum values.

6.6 Find the maximum value of the following function

$$f(x) = \frac{1}{x^2}$$

a) where $x \in D = [0, 1], \quad \mu_D(x) = 1.0$

b)   where $x \in D = [0, 1]$, $\mu_D(x) = x$

6.7 Integrate the function between A and B
$$f(x) = x^2+1$$
$$A = \{(1, 0.5), (2. 0.9)\} \quad B = \{(8. 0.1), (9. 0.5)\}$$

6.8 Differentiate the function at fuzzy point A
$$f(x) = 5\,x^2+1$$
$$A = \{(-2, 0.5), (-1, 0.9), (0, 1.0)\}$$

6.9 There is a fuzzy bunch of functions.
$$\tilde{f} = \{(f_1, 0.4), (f_2, 0.9), (f_3, 0.5), (f_4, 0.4)\}$$
$$x \in D = [1, 3]$$
$$f_1(x) = x+1$$
$$f_2(x) = x^2+1$$
$$f_3(x) = \frac{1}{x}$$
$$f_4(x) = \frac{1}{x} + 5$$
Find fuzzy sets $\tilde{f}(2)$ and $\tilde{f}(3)$.

6.10 Find fuzzy maximizing set M and maximum value of the following
    function.
$$f_1(x) = \sin x+1, x \in D$$
$$\mu_D(x) = \frac{x}{2\pi} \quad \text{for } 0 \le x \le 2\pi$$
$$= 0 \quad \text{otherwise}$$

6.11 There is a fuzzy function
$$\tilde{f} = \{(f_1, 0.4), (f_2, 0.5), (f_3, 0.4), (f_4, 0.9)\}, \quad x \in D = [1, 3]$$
$$f_1(x) = x+1$$
$$f_2(x) = x- 1$$
$$f_3(x) = x^2+1$$
$$f_4x) = x^2- 1$$
Find integration result of the function in $[1, 3]$.

6.12 There is a fuzzifying function
$$\tilde{f} = \{(f_1, 0.5), (f_2, 0.9), (f_3, 0.5)\},$$
$$f_1(x) = x^2+1$$
$$f_2(x) = x^3+ x^2+1$$
$$f_3(x) = x$$
Differentiate the function at $x_0 = 2$.

Chapter 7. PROBABILISY AND UNCERTAINTY

In this chapter, we will compare the fuzzy theory with the probability theory which is also used to express uncertainty. First, we briefly review the probability theory. The concept of fuzzy event is also described and then we study the characteristics of uncertainty. The concept of fuzziness is introduced in order to measure the uncertainty of fuzzy set.

7.1 Probability and Possibility

Since Zadeh proposed the fuzzy concept in 1965, there have been many discussions about the relationship between the fuzzy theory and probability theory. Both theories express uncertainty, have their values in the range of [0, 1], and have similarities in many aspects. In this section, we review the definitions of the two theories and compare them.

7.1.1 Probability Theory

Probability theory deals with the probability for an element to occur in universal set. We call the element as **event** and the set of possible events as sample space. In the sample space, the elements, i.e., events are **mutually exclusive**.

Example 7.1 When we play a six-side-dice, the sample space is $S=\{1, 2, 3, 4, 5, 6\}$. Among these six events, only one event can occur. The probability for any of these six events is 1/6.

An event might contain multiple elements. Consider two events A and B as follows.
$$A = \{1,3,5\}, \quad B = \{1,2,3\}.$$
The union and the intersection of these two events are,
$$A \cup B = \{1,2,3,5\}, \quad A \cap B = \{1,3\}$$
and the complement of the event A is
$$\overline{A} = \{2,4,6\}. \quad \square$$

172 7. Probability and Uncertainty

Definition (Probability distribution) To express the probability of events to occur in the sample space, we can define the ***probability distribution*** as follows.

The probability distrivution P is a numerically valued function that assigns a number $P(A)$ to event A so that the following axioms hold. In the axioms, S denotes the sample space.

i) $0 \leq P(A) \leq 1$,

ii) $P(S) = 1$,

iii) For the mutually exclusive events A_1, A_2, ... (that is, for any $i \neq j$, $A_i \cap A_j = \Phi$)

$$P(\bigcup_{i=1}^{\infty} A_i) = \sum_{i=1}^{\infty} P(A_i).$$ □

In the definition, $P(A)$ is the probability of an event A and the sample space S is the domain of probability distribution function. In the example of a six-side-dice, the probability P is,

$$P(i) = \frac{1}{6}, \quad i = 1,2,...,6.$$

If the probability distribution is defined, the following properties are satisfied.

(1) $P(A \cup B) = P(A) + P(B) - P(A \cap B)$.

(2) $P(A \cup B) = P(A) + P(B)$, if $A \cap B = \varnothing$.

(3) $P(A) + P(\overline{A}) = 1$.

Now assume that there are two sample spaces S and S', and an event A can occur in S and B in S'. When these events can occur in the mutually independent manner, the ***joint probability*** $P(AB)$ for both A and B to occur is

$$P(AB) = P(A) \cdot P(B).$$

The ***conditional probability*** $P(A \mid B)$ for A provided that the event B has occurred is

$$P(A \mid B) = \frac{P(AB)}{P(B)}.$$

7.1.2 Possibility Distribution

Fuzzy set A is defined on an universal set X and each element in the universal set has its membership degree in $[0, 1]$ for the set A.

$$\mu_A(x) > 0 \quad \text{for } x \in A$$
$$= 0 \quad \text{otherwise.}$$

The membership function μ_A can be defined as a possibility distribution function for the set A on the universal set X. The possibility of element x is denoted as $\mu_A(x)$ and these possibilities define the fuzzy set A.

We know the probability distribution P is defined on a sample space S and the sum of these probabilities should be equal to 1. Meanwhile, the possibility distribution is defined on an universal set X but there is not limit for the sum.

Example 7.2 Suppose the following proposition :

" Sophie has x sisters. "

$$x \in N = \{1, 2, 3, 4, \ldots 10\}.$$

Both probability distribution and possibility distribution can be used to define the variable x in N (Table 7.1). If we use the probability distribution P, the probability of having x sister(s) is defined by $P(x)$. By the possibility distribution μ, we define the possibility with x sisters as $\mu_A(x)$. The set N is considered as a sample space in the probability distribution and as a universal set in the possibility distribution.

Table 7.1. Possibility and Probability

x	1	2	3	4	5	6	7	8	9	10
P(x)	0.4	0.3	0.2	0.1	0	0	0	0	0	0
μ (x)	0.9	1.0	1.0	0.7	0.5	0.2	0.1	0	0	0

We see that the sum of the probabilitis is equel to 1 but that of the possibilities is greater than 1. In (Table 7.1.), we can see that higher possibility does not always means higher probability. But lower possibility leads to lower probability. So we can say that the possibility is the upper bound of the probability.   □

7.1.3 Comparison of Probability and Possibility

Probability and possibility have something in common: they both describe uncertainty.   The possibility can be regarded as the upper bound of probability value. That is, the possibility $\mu(A)$ and probability $P(A)$ of an event A have the following relation.

$$\mu(A) \geq P(A).$$

If the events A_1, A_2, ..., A_n are mutually exclusive, the probability of union of these events is equivalent to the sum of the probabilities of each event, and that of intersection is equivalent to the multiplication.

$$P(\cup_i A_i) = \sum_i P(A_i)$$

$$P(\cap_i A_i) = P(A_1) \cdot P(A_2) \cdot ... \cdot P(A_n).$$

The possibility for union of those events has the maximum value and that for intersection has the minimum. (Table 7.2) compares the characteristics of possibility with those of probability.

$$\mu(\cup_i A_i) = \underset{i}{\text{Max}}\, \mu(A_i)$$

$$\mu(\cap_i A_i) = \underset{i}{\text{Min}}\, \mu(A_i).$$

Table 7.2. Comparison of Possibility and Probability

	Possibility	Probability
Domain	Universal set X	Sample space S
Range	[0,1]	[0,1]
Constraints	none	$\sum_i P(A_i) = 1$
Union	$\mu(\cup_i A_i) = \underset{i}{\text{Max}}\, \mu(A_i)$	$P(\cup_i A_i) = \sum_i P(A_i)$
Intersection	$\mu(\cap_i A_i) = \underset{i}{\text{Min}}\, \mu(A_i)$	$P(\cap_i A_i) = P(A_1) \cdot P(A_2) \cdot ... \cdot P(A_n)$

7.2 Fuzzy Event

When dealing with the ordinary probability theory, an event has its precise boundary. For instance, if an event is $A=\{1,3,5\}$, its boundary is sharp and thus it can be represented as a crisp set. When we deal an event whose boundary is not sharp, it can be considered as a fuzzy set, that is, a fuzzy event. For example,

$$B = \text{`` small integer ''} = \{(1, 0.9), (2, 0.5), (3, 0.3)\}$$

How would we deal with the probability of such fuzzy events? We can identify the probability in two manners. One is dealing with the probability as a crisp value(crisp probability) and the other as a fuzzy set(fuzzy probability).

7.2.1 Crisp Probability of Fuzzy Event

Let a *crisp event* **A** be defined in the space $\Re^n$. All events in the space $\Re^n$ are mutually exclusive and the probability of each event is,

$$P(A) = \int_A dP$$

and for discrete event in the space $\Re^n$,

$$P(A) = \sum_{x \in A} P(x).$$

Let $\mu_A(x)$ be the *membership function* of the event(set) A and the *expectation* of $\mu_A(x)$ be $E_P(\mu_A)$. Then the following relationship is satisfied.

$$P(A) = \int_A \mu_A dP = E_P(\mu_A)$$

for discrete elements,

$$P(A) = \sum_{x \in A} \mu_A(x) P(x).$$

Definition (Crisp probability of fuzzy event) Let event A be a *fuzzy event* or a fuzzy set considered in the space $\Re^n$:

$$A = \{(x, \mu_A(x)) \mid x \in \Re^n\}$$

The probability for this fuzzy set is defined as follows:

$$P(A) = \int_A \mu_A dP = E_P(\mu_A)$$

and alternatively,

$$P(A) = \sum_{x \in A} \mu_A(x) P(x). \qquad \square$$

Example 7.3 Assume that the sample space $S=\{a,b,c,d\}$ is given as in (Fig 7.1.) Each element is mutually exclusive, and each probability is given as,

$$P(a) = 0.2,\ P(b) = 0.5,\ P(c) = 0.2,\ P(d) = 0.1.$$

Think about a crisp event $A=\{a,b,c\}$ in the sample space S with its characteristic function given as (Fig 7.2),

$$\mu_A(a) = \mu_A(b) = \mu_A(c) = 1,\ \mu_A(d) = 0$$

the probability of the crisp event A can be calculated from the following procedure.

$$P(A) = 1 \cdot 0.2 + 1 \cdot 0.5 + 1 \cdot 0.2 = 0.9.$$

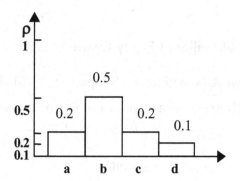

Fig. 7.1. Sample space S and Probability

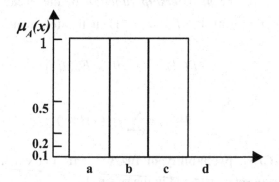

Fig. 7.2. Crisp event A

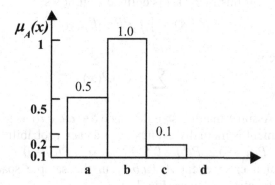

Fig. 7.3. Fuzzy event A

now consider the event A as a fuzzy event (Fig 7.3). That is,

$$A = \{(a,0.5),(b,1),(c,0.1)\}.$$

The crisp probability for the fuzzy event A is,

$$P(A) = 0.5 \cdot 0.2 + 1 \cdot 0.5 + 0.1 \cdot 0.2 = 0.62.$$

7.2.2 Fuzzy Probability of Fuzzy Event

Given the following fuzzy event in the sample space S,

$$A = \{(x, \mu_A(x)) \mid x \in S\}.$$

The α-cut set of the event(set) A is given as the following crisp set.

$$A_\alpha = \{x \mid \mu_A(x) \geq \alpha\}.$$

The probability of the α-cut event is as following:

$$P(A_\alpha) = \sum_{x \in A_\alpha} P(x).$$

Here, A_α is the union of mutually exclusive events. The probability of A_α is the sum of the probability of each event in the α-cut set A_α. For the probability of the α-cut event, we can say that

"The possibility of the probability of set A_α to be $P(A_\alpha)$ is α"

taking this interpretation, there are multiple cases for the fuzzy probability $P(A)$ according to the value α.

Definition (Fuzzy probability of fuzzy event) Fuzzy event A, its α-cut event A_α and the probability $P(A_\alpha)$ are provided from the above procedure. The fuzzy probability $P(A)$ is defined as follows:

$$P(A) = \{(P(A_\alpha),\alpha) \mid \alpha \in [0,1]\}. \quad \square$$

Of course, the value of α is an element in the level set of fuzzy set A. We used the same kind of interpretation when we discussed about the fuzzy cardinality of fuzzy set in chapter 1.

Example 7.4 Assume the probability of each element in the sample space $S=\{a,b,c,d\}$ as shown in (Fig 7.4.)

$$P(a) = 0.2,\ P(b) = 0.3,\ P(c) = 0.4,\ P(d) = 0.1$$

A fuzzy event A is given in (Fig 7.5.)

$$A = \{(a,1),(b,0.8),(c,0.5),(d,0.3)\}$$

taking the α-cut event A_α, we get crisp events.

$$A_{0.3} = \{a,b,c,d\}$$

$$A_{0.5} = \{a,b,c\}$$

$$A_{0.8} = \{a,b\}.$$

$$A_1 = \{a\}$$

since these are crisp events, we can easily calculate the probabilities of each α-cut event.

$$P(A_{0.3}) = 0.2 + 0.3 + 0.4 + 0.1 = 1$$
$$P(A_{0.5}) = 0.2 + 0.3 + 0.4 = 0.9$$
$$P(A_{0.8}) = 0.2 + 0.3 = 0.5$$
$$P(A_1) = 0.2 .$$

Now, the possibility for this fuzzy event A to be $P(A_\alpha)$ is α, and the probability of the fuzzy event A is given as follows (Fig7.6):

$$P(A) = \{(1,0.3),(0.9,0.5),(0.5,0.8),(0.2,1)\} . \qquad \square$$

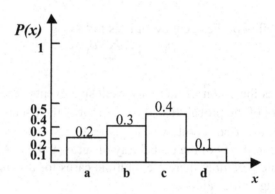

Fig. 7.4. Sample space $S = \{a,b,c,d\}$

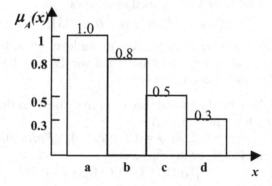

Fig. 7.5. Fuzzy event A

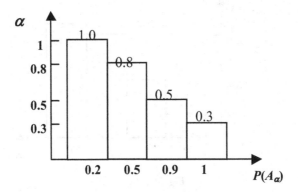

Fig. 7.6. Fuzzy probability

7.3 Uncertainty

7.3.1 Uncertainty Level of Element

Suppose students *a, b, c* and *d* take the entrance examination for college *A*. The possibility for each student to enter the college is 0.2, 0.5, 0.9 and 1 respectively. At this point, the possibilities can be identified as a fuzzy set. That is, let the college be a fuzzy set *A* and *a, b, c* and *d* be the elements of *A*. The possibilities to be contained in *A* can be expressed as the values of membership function (Fig 7.7).

$$\mu_A(a) = 0.2, \quad \mu_A(b) = 0.5, \quad \mu_A(c) = 0.9, \quad \mu_A(d) = 1.$$

When discussing the possibilities to be in *A*, which of these elements *a, b, c* and *d* has the largest uncertainty?

First, the element *d* has the concrete possibility, we say it has the least uncertainty. The element *a*, on the other hand, has almost no possibility to pass the examination. This student doesn't expect too much for his success and we might say he has relatively less uncertainty. However, the student *b* might have the most uncertainty since he has the possibility 0.5.

When the possibility is near to 0.5, the uncertainty gets the highest. As the possibility approaches 0 or 1, the uncertainty decreases.

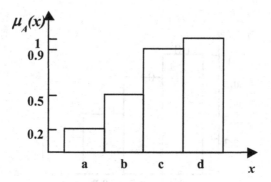

Fig. 7.7. Fuzzy set *A*

7.3.2 Fuzziness of Fuzzy Set

Suppose students *a, b, c* and *d* apply to *B* college and consider *B* as a fuzzy set, the possibilities for those students to succeed in the entrance examination are shown in the following membership functions (Fig 7.8).

$$\mu_B(a) = 0.4, \quad \mu_B(b) = 0.7, \quad \mu_B(c) = 0.6, \quad \mu_B(d) = 0.5.$$

Comparing the fuzzy sets A and B, which one is more uncertain? Comparisons element by element, the elements of *B* are more uncertain (the values of membership functions are closer to 0.5). So, the fuzzy set *B* has more uncertain states comparing with *A*. So as to speak, *B* is relatively more fuzzy.

If we consider another fuzzy set *C*, each element of which having its membership degree 0.5, the fuzzy set *C* has the largest degree of fuzziness (uncertainty).

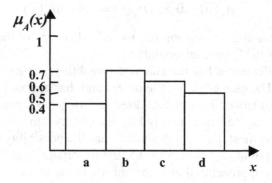

Fig. 7.8. Fuzzy set B

7.4 Measure of Fuzziness

7.4.1 Definition

Fuzziness is termed for the case when representation of uncertainty level is needed. The function for this fuzziness is called ***measure of fuzziness***. The function f denoting the measure of fuzziness is

$$f : P(X) \to \text{R}.$$

In the function, $P(X)$ is the power set gathering all subsets of the universal set X, and R is the real number domain. The function f grants the real value $f(A)$ to the subset A of X, and the value indicates the fuzziness of set A, there are three conditions for the measures to observe.

(Axiom F1) $f(A) = 0$ iff A is a crisp set.

The fuzziness should have the value 0 when the set is crisp (Fig 7.9).

(Axiom F2) When the uncertainty of A is less than that of B , the measured value $f(A)$ should be less than $f(B)$. When we denote it by $A < B$ and say that A is sharper than B, the following relation should be satisfied.

$$f(A) \leq f(B).$$

The relation implies the ***monotonicity*** property.

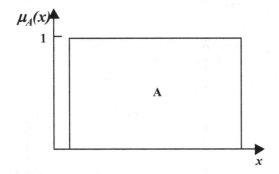

Fig. 7.9. Fuzziness of crisp set $f(A) = 0$ (minimum uncertainty).

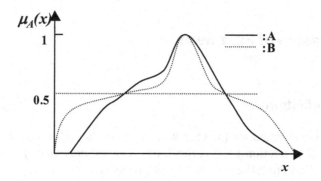

Fig. 7.10. When A is sharper than B, $f(A) \leq f(B)$.

Example 7.5 When two fuzzy sets A and B are presented (Fig 7.10), the fact that A is sharper than B is defined as following. That is, if $A < B$ holds, then the followings are satisfied.

i)   when $\mu(x) \leq \frac{1}{2}$, $\mu_A(x) \leq \mu_B(x)$

ii) when $\mu(x) \geq \frac{1}{2}$, $\mu_A(x) \geq \mu_B(x)$

By the axiom F2, the relation $f(A) \leq f(B)$ should be satisfied.   □

(Axiom F3) If the fuzziness is the maximum, the measure $f(B)$ should have the maximum value.

Example 7.6 For the deep understanding of the axiom F3, think of the case that the uncertainty is the maximum. If the membership degree of each element x in fuzzy set A is 0.5, the uncertainty has the maximum. That is, for all elements $x \in A$, $\mu_A(x) = 0.5$, and then the fuzziness measure $f(A)$ is also the maximum (Fig 7.11).   □

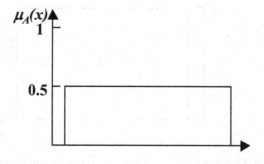

Fig. 7.11. Maximum uncertainty $f(A)$.

7.4.2 Measure using Entropy

In this section, a fuzziness measure based on Shannon's entropy is explained. Shannon's entropy is widely used in measuring the amount of uncertainty or information, and is considered as the fundamental theory in the information theory.

Definition (Shannon's entropy)

$$H(P(x)) = -\sum_{x \in X} P(x) \log_2 P(x), \quad \forall x \in X. \quad \square$$

In the rest of the section, $P(x)$ denotes the probability distribution in the universal set X for all $x \in X$.

Example 7.7 To know more about the entropy, consider the following examples with two probability distributions P and P'.

i) Probability distribution P

For the universal set $X=\{a,b,c\}$, the probability distribution P is given as,

$$P(a) = 1/3, \quad P(b) = 1/3, \quad P(c) = 1/3,$$

$$P(a) + P(b) + P(c) = 1.$$

The probabilities of all elements in X are equal to each other, and the sum of the probabilities is 1. The uncertainty of one element's occurrence is measured by the Shannon's entropy.

$$H(P) = -\left[\tfrac{1}{3}\log_2 \tfrac{1}{3} + \tfrac{1}{3}\log_2 \tfrac{1}{3} + \tfrac{1}{3}\log_2 \tfrac{1}{3}\right]$$

$$= -\log_2 \tfrac{1}{3} = \log_2 3 = 1.6.$$

ii) Probability distribution P'

Assume there is a probability distribution P' for X as follows.

$$P'(a) = 1/2, \quad P'(b) = 1/2, \quad P'(c) = 0$$

The uncertainty is,

$$H(P') = -\left[\tfrac{1}{2}\log_2 \tfrac{1}{2} + \tfrac{1}{2}\log_2 \tfrac{1}{2} + 0\right]$$

$$= -\log_2 \tfrac{1}{2} = \log_2 2 = 1.$$

The uncertainty for P is greater than that of P' (Fig 7.12). $\square$

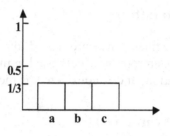

probability distribution P

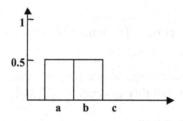

probability distribution P'

$$H(P) = 1.6 > 1 = H(P')$$

Fig. 7.12. Shannon's entropy in probability distribution

In the above example, when the probability distribution is P', the possibility for the element c to occur is 0. So we do not need to consider c. But in the probability distribution P, we need to consider the three elements, a,b,c. So the uncertainty $H(P)$ is greater than $H(P')$. When there are only two events and the probability of each event is 0.5, the amount of information is the maximum at 1 as in the case P'

Since the Shannon's entropy is based on the probability distribution, the total probabilities of all elements is 1.

$$\sum_X P(x) = 1.$$

But for fuzzy sets, this restriction is unnecessary. If a fuzzy set A is defined in the universal set X by a membership function $\mu_A(x)$, the following restriction is not required.

$$\sum_X \mu_A(x) = 1 \quad \text{(not necessary)}.$$

Now referring to what we have seen, define a measure of fuzziness $f(A)$ of a fuzzy set A.

Definition (measure of fuzziness)

$$f(A) = -\sum_{x \in X} \left[\mu_A(x) \log_2 \mu_A(x) + (1 - \mu_A(x)) \log_2 (1 - \mu_A(x)) \right]. \quad \square$$

This is the sum of the uncertainties of a fuzzy set A defined by the membership function $\mu_A(x)$ and its complement $\overline{A}$ defined by$[1-\mu_A(x)]$

The normalized measure $\hat{f}(A)$ of the measure $f(A)$ is defined as the following

$$\hat{f}(A) = \frac{f(A)}{|X|}.$$

In the above,$|X|$ denotes the cardinality of the universal set X and the normalized measure observes this relation

$$0 \le \hat{f}(A) \le 1$$

this measure satisfies the axiom (F1) and (F2).

Example 7.8 Suppose that there are two fuzzy sets A and A' in $X=\{a,b,c\}$ (Fig 7.13).

i) Assume that a fuzzy set A is given as
$$A = \{(a,0.5),(b,0.2),(c,1)\}.$$

The fuzziness of the fuzzy set A is,

$$f(A) = -(0.5\log 0.5 + 0.5\log 0.5 + 0.2\log 0.2 + 0.8\log 0.8 + 1\log 1 + 0)$$

$$= -(\log_2 \frac{1}{2} + \frac{1}{5}\log_2 \frac{1}{5} + \frac{4}{5}\log_2 \frac{4}{5})$$

$$= \log_2 2 + \frac{1}{5}\log_2 5 + \frac{4}{5}\log_2 \frac{5}{4}$$

$$= \log_2 5 - 0.6 = 1.7 .$$

and the normalized measure yields,

$$\hat{f}(A) = \frac{f(A)}{|X|} = \frac{1.7}{3} = 0.57 .$$

ii) The fuzzy set A' is given as follows.

$$A' = \{(a,0.5),(b,0.5),(c,0.5)\}$$

The fuzziness for the fuzzy set A' is,

$$f(A') = -(0.5\log_2 0.5 + 0.5\log_2 0.5 + 0.5\log_2 0.5$$
$$+ 0.5\log_2 0.5 + 0.5\log_2 0.5 + 0.5\log_2 0.5)$$
$$= -(3\log_2 0.5) = 3\log_2 2 = 3 ,$$

and the normalized measure is,

$$\hat{f}(A) = \frac{f(A')}{|X|} = \frac{3}{3} = 1 .$$

The membership degrees of all elements in A' are 0.5. So the uncertainty of the fuzzy set A' is larger. Consequently, the fuzziness of A' is greater than that of A

$$f(A) < f(A'). \qquad \square$$

We state that the uncertainty is the largest when the membership degrees are all 0.5. And the normalized fuzziness of such fuzzy set is 1.

7.4.3 Measure using Metric Distance

Another measure of fuzziness is the one that is based on the concept of metric distance. We talked about Hamming distance and Euclidean distance in sec 2.5. A crisp set C that corresponds to a fuzzy set A is introduced for the distance measure (Fig 7.14).

$$\mu_C(x) = 0 \qquad \text{if } \mu_A(x) \le \frac{1}{2}$$

$$\mu_C(x) = 1 \qquad \text{if } \mu_A(x) > \frac{1}{2}$$

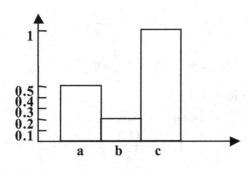

fuzzy set A

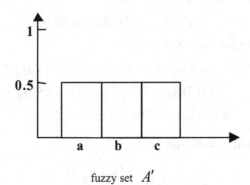

fuzzy set A'

Fig. 7.13. Measure of fuzziness for A and A'

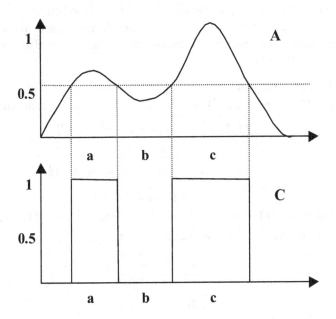

Fig. 7.14. Fuzzy set A and its corresponding Crisp set C

If the set C is defined as such, the measure of fuzziness is the distance between the fuzzy set A and the crisp set C. For the measure of distance, we can use Hamming distance or Euclidean distance (see sec 2.5.3).

Definition (Hamming distance) The measure of fuzziness $f(A)$ is expressed as,

$$f(A) = \sum_{x \in X} |\mu_A(x) - \mu_C(x)| . \qquad \square$$

Definition (Euclidean distance) The measure of fuzziness $f(A)$ is,

$$f(A) = \left(\sum_{x \in X} [\mu_A(x) - \mu_C(x)]^2 \right)^{\frac{1}{2}} . \qquad \square$$

The closer to 0.5 the values of membership function are, the larger the fuzziness is measured.

Definition (Minkowski distance) Generalizing Hamming distance and Euclidean distance, the following ***Minkowski's measure*** yields

$$f_w(A) = \left(\sum_{x \in X} |\mu_A(x) - \mu_C(x)|^w \right)^{\frac{1}{w}} . \qquad \square$$

The Minkowski's measure holds for $w \in [1, \infty]$. When $w=1$, it becomes Hamming distance, and when $w=2$, it is Euclidean distance. We easily see that the previously introduced measures of fuzziness also satisfy the axioms (F1) and (F2).

Example 7.9 Consider the sets A and A' in (Fig 7.13.) Hamming distances $f(A)$ and $f(A')$ can be calculated as in the following.

$$f(A) = |0.5 - 0| + |0.2 - 0| + |1 - 1|$$
$$= 0.5 + 0.2 + 0 = 0.7$$
$$f(A') = |0.5 - 0| + |0.5 - 0| + |0.5 - 0|$$
$$= 0.5 + 0.5 + 0.5 = 1.5.$$

The relation $f(A) < f(A')$ holds in the above. The Hamming measure of fuzziness can be normalized like this.

$$\hat{f}(A) = \frac{f(A)}{0.5|X|}$$

$$0 \leq \hat{f}(A) \leq 1.$$

The symbol $|X|$ denotes the cardinality of the universal set X. The normalized measure of the previous example is

$$\hat{f}(A) = \frac{0.7}{0.5 \times 3} = 0.47. \qquad \square$$

[SUMMARY]

☐ Probability theory
  – Sample space, event, mutually exclusive event
  – Conditional probability : $P(A\,|\,B)$
  – Joint probability : $P(AB)$

☐ Discrimination of possibility and probability
  – Restriction : whether $\sum P(A)=1$ or not
  – Set operations of union and intersection : Max, Min operation

☐ Fuzzy event
  – Crisp probability
  – Fuzzy probability

☐ Conditions for the measure of the degree of fuzziness
  – In a crisp set A, $f(A)=0$
  – Monotonicity
  – In the maximum fuzziness, $f(A)$ is the maximum

☐ Measure of the fuzziness using entropy
$$f(A) = \sum_{x \in X}\left(\mu_A(x)\log_2 \mu_A(x) + \left[1 - \mu_A(x)\right]\log_2\left[1 - \mu_A(x)\right]\right).$$

☐ Measure of the fuzziness using metric distance
  – Hamming distance
$$f(A) = \sum_{x \in X}\left|\mu_A(x) - \mu_C(x)\right|.$$
  – Eclidiean distance
$$f(A) = \left(\sum_{x \in X}\left[\mu_A(x) - \mu_C(x)\right]^2\right)^{\frac{1}{2}}.$$

[EXERCISES]

7.1 Show operations of the following formulas in the possibility theory
   a) $\mu(\bigcup_i A_i)$

   b) $\mu(\bigcap_i A_i)$

7.2 Show operations of the formulas in the probability theory.
   a) $p(\bigcup_i A_i)$

   b) $p(\bigcap_i A_i)$

7.3 There is a fuzzy event A
$$A = \{(1, 0.8), (2, 0.8), (3, 0.5), (4, 0.3), (5. 0.1)\}$$
Apply α-cut operations to the fuzzy event.

7.4 There is a probability distribution $P(x)$, and two events A and B.
$$P(a) = 0.2, P(b)=0.5, P(c)=0.2, P(d)=0.1$$
$$A = \{a, b, c\}$$
$$B = \{(a, 0.5), (b, 0.9)m (c, 0.7), (d, 0.1)\}$$
   a) Find the probability of event A
   b) Fine the crisp probability of fuzzy event B
   c) Find the fuzzy probability of fuzzy event B

7.5 Consider a probability distribution $P(x)$ on an sample space $S = \{1, 2, 3, 4, 5, 6\}$.
$P(1) = 0.2, P(2) = 0.1, P(3) = 0.1), P(4) = 0.2, P(5) = 0.1 , P(6) = 0.3$
Compare probabilities of the following two fuzzy events :
$$A = \{ (1, 0.5), (2, 0.5), (3, 0.5), (4, 0.5)\}$$
$$B = \{(1, 0.9), (2, 0.1), (3, 0.2), (4, 0.8)\}$$

7.6 consider a probability distribution $P(x)$ on a sample space
   $S = \{a, b, c, d\}$
$$p(a) = \frac{1}{4}, \quad p(b) = \frac{1}{4}, \quad p(c) = \frac{1}{4}, \quad p(d) = \frac{1}{4}$$
Find the Shannan's entropy in the probability distribution.

7.7 There is a fuzzy set A defined on the universal set
   $X = \{a, b, c, d\}$
   $A = \{(a, 0.5), (b, 0.2), (c, 0.8), (d, 0.1)\}$

Find the fuzziness of the set A by using the concept of Shannon 's entropy.

7.8 There is a fuzzy set $A = \{(1, 0.4), (2, 0.8), (3, 0.4), (4, 0.1)\}$
   a)   Fined the fuzziness of A using Hanming distance
   b)   Fined the fuzziness of A using Euclidean distance

Chapter 8. FUZZY LOGIC

Formal language is a language in which the syntax is precisely given and thus is different from informal language like English and French. The study of the formal languages is the content of mathematics known as mathematical logic. The mathematical logic is called classical logic in this chapter. The classical logic considers the binary logic which consists of truth and false. The fuzzy logic is a generalization of the classical logic and deals with the ambiguity in the logic. In this chapter, we summarize the classical logic and then study the fuzzy logic.

8.1 Classical Logic

8.1.1 Proposition Logic

Definition (Proposition) As in our ordinary informal language, "sentence" is used in the logic. Especially, a sentence having only "true (1)" or "false (0)" as its truth value is called "proposition".

Example 8.1 The following sentences are propositions.

Smith hits 30 home runs in one season.	(true)
$2 + 4 = 7$	(false)
For every x, if $f(x) = \sin x$, then $f'(x) = \cos x$.	(true)
It rains now.	(true) □

Example 8.2. The followings are not propositions.

Why are you interested in the fuzzy theory?
He hits 5 home runs in one season.
$$x + 5 = 0$$
$$x + y = z$$

In the second example, we do not know who is "He" and thus cannot determine whether the sentence is true (1) or false (0). If "He" is replaced by "Tom", we have

Tom hits 5 home runs in one season.

Now we can evaluate the truth value of the above sentence. In the same way,

$$x + 5 = 0$$

is not a proposition.   If the variable x is replaced by –5, then the sentence

$$-5 + 5 = 0$$

has its value "true".   A variable is used as a symbol representing an element in a universal set.   □

Definition (Logic variable) As we know now, a proposition has its value (true or false).   If we represent a proposition as a variable, the variable can have the value true or false.   This type of variable is called as a "proposition variable" or "logic variable".   □

We can combine prepositional variables by using "connectives".   The basic connectives are negation, conjunction, disjunction, and implication.

(1)   Negation
Let's assume that prepositional variable P represents the following sentence.

$$P: 2 \text{ is rational.}$$

In this case, the true value of P is true.

$$P = \text{true}$$

but its negation is false and represents as follows.

$$\overline{P} = \text{false}$$

The truth table representing the values of negation is given in (Table 8.1.)

(2)   Conjunction
If a and b are prepositional variables, their conjunction is represented as follows and is interpreted as "a AND b".

$$a \wedge b$$

The truth value of the above conjunction is determined according to the values of a and b (Table 8.2).

Example 8.3 Suppose there are two propositions a and b. We can see their conjunction is 0.

$$a: 2 + 2 = 4$$
$$b: 3 + 2 = 7$$
$$c: a \wedge b$$

then, a = 1, b = 0, and c = 0.   □

Table 8.1. Truth table of negation

a	$\overline{a}$
1	0
0	1

Table 8.2. Truth table of conjunction

a	b	a ∧ b
1	1	1
1	0	0
0	1	0
0	0	0

(3) Disjunction

The disjunction of two propositions a and b is represented as follows

$$a \vee b$$

The disjunction is interpreted as "a OR b". But it has two different meanings: "exclusive OR" and "inclusive OR". The exclusive OR is used in which two events could not happen simultaneously.

Are you awake or asleep?

The inclusive OR is used when two events can occur simultaneously.

Are you wearing a shirt or sweater?

in general, if we say the disjunction, we mean the "inclusive OR" (Table 8.3).

(4) Implication

The proposition "if a, then b." is represented as follows.

$$a \rightarrow b$$

Example 8.4 Consider the following propositions. We study the truth value of each proposition in varying the value of propositional variables a and b.

Table 8.3. Truth table of disjunction

a	b	a ∨ b
1	1	1
1	0	1
0	1	1
0	0	0

Table 8.4. Truth table of implication

a	b	a → b
1	1	1
1	0	0
0	1	1
0	0	1

i)   a → b where a: 2 + 2 = 4, b: 3 + 3 = 6
ii) a → b where a: 2 + 2 = 4, b: 3+3 = 7
iii) a → b where a: 2 + 2 = 5, b: 3 + 3 = 6
iv) a → b where a: 2 + 2 = 5, b: 3 + 3 = 7

we can see that the above propositions are true except for the second.   □

The truth values of implication are summarized in (Table 8.4.)   In the table we see that the value of implication can be represented by $\bar{a} \vee b$.

8.1.2 Logic Function

The "logic function" is a combination of propositional variables by using connectives.   Values of the logic function can be evaluated according to the values of propositional variables and the truth values of connectives. As we know, a logic function having only one propositional variable has two kinds of values: true and false.   A logic function containing two variables has 4 ($=2^2$) different combinations of values: (true, true), (true, false), (false, true), (false, false).   Similarly a function having n variables can have 2^n different combinations.

If a logic function has one or two prepositional (logic) variables, it is called a "logic primitive".   By using the logic primitive, we can represent an (algebraic) expression which is called a "logic formula".

Definition (Logic formula) The logic formula is defined as following :

i)   Truth values 0 and 1 are logic formulas

ii) If v is a logic variable, v and $\bar{v}$ are a logic formulas

iii) If a and b represent a logic formulas, a ∧ b and a ∨ b are also logic
       formulas.

iv) The expressions defined by the above (1), (2), and (3) are logic
       formulas.   □

Table 8.5. Properties of classical logic

(1) Involution	$\overline{\overline{a}} = a$
(2) Commutativity	$a \wedge b = b \wedge a$
	$a \vee b = b \vee a$
(3) Associativity	$(a \wedge b) \wedge c = a \wedge (b \wedge c)$
	$(a \vee b) \vee c = a \vee (b \vee c)$
(4) Distributivity	$a \vee (b \wedge c) = (a \vee b) \wedge (a \vee c)$
	$a \wedge (b \vee c) = (a \wedge b) \vee (a \wedge b)$
(5) Idempotency	$a \wedge a = a$
	$a \vee a = a$
(6) Absorption	$a \vee (a \wedge b) = a$
	$a \wedge (a \vee b) = a$
(7) Absorption by 0 and 1	$a \wedge 0 = 0$
	$a \vee 1 = 1$
(8) Identity	$a \wedge 1 = a$
	$a \vee 0 = a$
(9) De Morgan's law	$\overline{a \wedge b} = \overline{a} \vee \overline{b}$
	$\overline{a \vee b} = \overline{a} \wedge \overline{b}$
(10) Absorption of complement	$a \vee (\overline{a} \wedge b) = a \vee b$
	$a \wedge (\overline{a} \vee b) = a \wedge b$
(11) Law of contradiction	$a \wedge \overline{a} = 0$
(12) Law of excluded middle	$a \vee \overline{a} = 1$

Any logic formula defines a logic function, and it has its truth value. Properties of logic formulas are summarized in (Table 8.5.)

Some of important logic formulas and their values are given in the following:

(1)   Negation $\overline{a} = 1 - a$

(2)   Conjunction $a \wedge b = $ Min (a, b)

(3) Disjunction $a \vee b = \text{Max}\,(a, b)$
(4) Implication $a \rightarrow b = \bar{a} \vee b$

8.1.3 Tautology and Inference Rule

Definition (Tautology) A "tautology" is a logic formula whose value is always true regardless of its logic variables. A "contradiction" is one which is always false.   □

Example 8.5 Consider the following logic formula.

$$\overline{(a \rightarrow b)} \rightarrow \bar{b}$$

This proposition means that if the value of $(a \rightarrow b)$ is false then b is false.   Let's evaluate its value with different values of logic variables a and b in (Table 8.6.)

Table 8.6. Truth value of tautology $\overline{(a \rightarrow b)} \rightarrow \bar{b}$

a	b	$a \rightarrow b$	$\overline{(a \rightarrow b)}$	$\bar{b}$	$\overline{(a \rightarrow b)} \rightarrow \bar{b}$
1	1	1	0	0	1
1	0	0	1	1	1
0	1	1	0	0	1
0	0	1	0	1	1

in Table 8.6, we see that the logic formula is always true and thus it is a tautology.   That is, if $\overline{(a \rightarrow b)}$ is true, then $\bar{b}$ is always true.   □

Example 8.6 Let's consider another example.

$$(a \wedge (a \rightarrow b)) \rightarrow b$$

The truth values of this proposition are evaluated in (Table 8.7.)
we can see that this proposition has also true value regardless of the values of a and b. This tautology means that
"If a is true and $(a \rightarrow b)$ is true, then b is true." or
"If a exists and the relation $(a \rightarrow b)$ is true, then b exists."

Table 8.7. Truth value of tautology $(a \wedge (a \rightarrow b)) \rightarrow b$

a	b	$(a \rightarrow b)$	$(a \wedge (a \rightarrow b))$	$(a \wedge (a \rightarrow b)) \rightarrow b$
1	1	1	1	1
1	0	0	0	1
0	1	1	0	1
0	0	1	0	1

The interpretations in the above example show a logic procedure of tautology, and thus we can obtain a correct conclusion when we follow the logic procedure. Therefore, the tautology is used as a rule of "deductive inference". There are some important inference rules using tautologies.

$$(a \wedge (a \rightarrow b)) \rightarrow b \qquad \text{: modus ponens}$$
$$(\overline{b} \wedge (a \rightarrow b)) \rightarrow \overline{a} \qquad \text{: modus tollens}$$
$$((a \rightarrow b) \wedge (b \rightarrow c)) \rightarrow (a \rightarrow c) \qquad \text{: hypothetical syllogism}$$

8.1.4 Predicate Logic

A "predicate" is a group of words like

"is a man"
"is green"
"is less than"
"belongs to"

They can be applied to one or more names of individuals (objects) to yield meaningful sentences; for example,

"Socrates is a man."
"Two is less than four."
"That hat belongs to me."
"He is John."

the names of the individuals are called individual constants.

Definition (Predicate logic) "Predicate logic" is a logic which represents a proposition with the predicate and individual (object).   □

Example 8.7 The following propositions are "predicate propositions" and consist of predicates and objects.

"Socrates is a man"
        predicate: "is a man"
        object: "Socrates"
"Two is less than four"
        predicate: "is less than"
        object: "two" , "four"

Sometimes, the objects can be represented by "variable", and then, in that case, the "predicate proposition" can be evaluated if an element in the universal set is instantiated to the variable.   □

Example8.8 Let's consider the following examples.

i) x is a man.
ii) y is green.
iii) z is less than w.
iv) p belongs to q.

If an individual is mapped to a variable, the sentence can have its meaning and then we can evaluate the value of proposition. There are examples of individual constants for the above propositions with variables.

(1) <u>Tom</u> is a man.
(2) <u>His face</u> is green.
(3) <u>Two</u> is less than <u>four</u>.
(4) <u>This hat</u> belongs to <u>me</u>.

In the first sentence, Tom is an element in the universal set "man". The element Tom is instantiated to the variable x, and then we know the value of the proposition is true. In the second sentence, his face is an object (element) corresponded to the variable y, and we can evaluate the truth value of the proposition.   □

In the formal language of this chapter, we denote predicates by letters. For example, the sentence "x satisfies P" can be written by P(x).

Example 8.9 For example, the above predicate propositions can be represented in the following way.

      is_a_man(x),   is_a_man(Tom)
      is_green(y),   is_green(his face)
      is_less_than(z, w),   is_less_than(two, four)
      belongs_to(p, q),   belongs_to(this hat, me)   □

The number of individual constants to which a given predicate is called number of places of the predicate. For instance, "is a man" is a one-place predicate, and "is less than" is a two-place predicate.

A one-place predicate determines a set of things: namely those things for which it is true. Similarly, a two-place predicate determines a set of pairs of things; that is, a two-place "relation". In general, an n-place predicate determines an n-place relation. We may think of the predicate as denoting the relation.

Example 8.10 For example, the predicate "is man" determines the set of men, and the predicated "is south of" determines the set of pairs (x, y) of cities such that x is south of y. For instance, the relation holds when x = Sydney and y = Tokyo, but not when x = New York and y = Seoul. Different predicates may determine the same relation. For example, "x is south of y" and "y is north of x".   □

8.1.5 Quantifier

The phrase "for all" is called the "universal quantifier" and is denoted symbolically by ∀. The phrase "there exists", "there is a", or "for some" is called the "existential quantifier" and is denoted symbolically by ∃.

The universal quantifier is kind of an iterated conjunction. Suppose there are only finitely-many individuals. That is, the variable *x* takes only the values $a_1, a_2, \ldots a_n$. Then the sentence $\forall x P(x)$ has the same meaning as the conjunction $P(a_1) \wedge P(a_2) \wedge \ldots \wedge P(a_n)$.

The existential quantifier is kind of an iterated disjunction. If there are only finitely-many individuals $a_1, a_2, \ldots a_n$, then the sentence $\exists x P(x)$ has the same meaning as the disjunction $P(a_1) \vee P(a_2) \vee \ldots \vee P(a_n)$.

Of course, if the number of individuals is infinite, such an interpretation of the quantifier is not possible, since infinitely long sentences are not allowed.

According to De Morgan's laws, $P(a_1) \vee P(a_2) \vee \ldots \vee P(a_n)$ is equivalent to $\sim [\sim P(a_1) \wedge \sim P(a_2) \wedge \ldots \wedge \sim P(a_n)]$ where the symbol $\sim$ represents the negative operator. This suggests the possibility of defining the existential quantifier from the universal quantifier. We shall do this; $\exists x P(x)$ will be an abbreviation for $\sim \forall x \sim P(x)$. Of course we could also define the universal quantifier from the existential quantifier; $\forall x P(x)$ has the same meaning as $\sim \exists x \sim P(x)$.

8.2 Fuzzy Logic

8.2.1 Fuzzy Expression

In the fuzzy expression(formula), a fuzzy proposition can have its truth value in the interval $[0,1]$. The fuzzy expression function is a mapping function from $[0,1]$ to $[0,1]$.

$$f : [0,1] \to [0,1]$$

If we generalize the domain in n-dimension, the function becomes as follows:

$$f : [0,1]^n \to [0,1]$$

Therefore we can interpret the fuzzy expression as an n-ary relation from n fuzzy sets to $[0,1]$. In the fuzzy logic, the operations such as negation ($\sim$ or $\neg$), conjunction ($\wedge$) and disjunction ($\vee$) are used as in the classical logic.

Definition (Fuzzy logic) Then the fuzzy logic is a logic represented by the fuzzy expression (formula) which satisfies the followings.

  i) Truth values, 0 and 1, and variable x_i ($\in [0,1]$, i = 1, 2, ..., n) are fuzzy expressions.

  ii) If f is a fuzzy expression, $\sim$f is also a fuzzy expression.

  iii) If f and g are fuzzy expressions, f $\wedge$ g and f $\vee$ g are also fuzzy expressions.   □

8.2.2 Operators in Fuzzy Expression

There are some operators in the fuzzy expression such as $\neg$ (negation), $\wedge$ (conjunction), $\vee$ (disjunction), and $\rightarrow$ (implication). However the meaning of operators may be different according to the literature. If we follow Lukasiewicz's definition, the operators are defined as follows for a, b $\in [0,1]$.

(1) Negation $\bar{a} = 1 - a$
(2) Conjunction $a \wedge b = \text{Min}\,(a, b)$
(3) Disjunction $a \vee b = \text{Max}\,(a, b)$
(4) Implication $a \rightarrow b = \text{Min}\,(1, 1{+}b{-}a)$

The properties of fuzzy operators are summarized in (Table 8.9.)

Table 8.9. The properties of fuzzy logic operators

(1) Involution	$\bar{\bar{a}} = a$
(2) Commutativity	$a \wedge b = b \wedge a$
	$a \vee b = b \vee a$
(3) Associativity	$(a \wedge b) \wedge c = a \wedge (b \wedge c)$
	$(a \vee b) \vee c = a \vee (b \vee c)$
(4) Distributivity	$a \vee (b \wedge c) = (a \vee b) \wedge (a \vee c)$
	$a \wedge (b \vee c) = (a \wedge b) \vee (a \wedge c)$
(5) Idempotency	$a \wedge a = a$
	$a \vee a = a$
(6) Absorption	$a \vee (a \wedge b) = a$
	$a \wedge (a \vee b) = a$
(7) Absorption by 0 and 1	$a \wedge 0 = 0$
	$a \vee 1 = 1$
(8) Identity	$a \wedge 1 = a$
	$a \vee 0 = a$
(9) De Morgan's law	$\overline{a \wedge b} = \bar{a} \vee \bar{b}$
	$\overline{a \vee b} = \bar{a} \wedge \bar{b}$

But we have to notice that the law of contradiction and law of excluded middle are not verified in the fuzzy logic.

Example 8.11 We can see that the two properties are not satisfied in the following examples.

i) Law of contradiction

Assume a is in [0,1].

$$a \wedge \bar{a} = \text{Min}[a, \ \bar{a}] = \text{Min}[a, 1-a]$$
$$= \begin{cases} a & \text{if } 0 \le a \le 0.5 \\ 1-a & \text{if } 0.5 \le a \prec 1 \end{cases}$$
$$\therefore \ 0 < a \wedge \bar{a} \le 0.5$$
$$\text{Then } a \wedge \bar{a} \neq 0$$

ii) Law of excluded middle

Suppose a is in [0,1].

$$a \vee \bar{a} = \text{Max}[a, \ \bar{a}] = \text{Max}[a, 1-a]$$
$$= \begin{cases} a & \text{if } 0.5 \le a \prec 1 \\ 1-a & \text{if } 0 \prec a \le 0.5 \end{cases}$$
$$\therefore \ 0.5 \le a \vee \bar{a} < 1$$
$$\text{Then } a \vee \bar{a} = 1 \qquad \text{if a} = 0 \text{ or } 1$$
$$a \vee \bar{a} < 1 \quad \text{otherwise} \qquad \square$$

8.2.3 Some Examples of Fuzzy Logic Operations

In this section, we have two examples of classical logic operation and one example of fuzzy logic operation. In these examples, we will see that the fuzzy logic operation is a generalization of the classical one.

Example 8.12 When $a = 1, b = 0$
  i) $\bar{a} = 0$
  ii) $a \wedge b = \text{Min}(1, 0) = 0$
  iii) $a \vee b = \text{Max}(1, 0) = 1$
  iv) $a \rightarrow b = \text{Min}(1, 1-1+0) = 0$ $\square$

Example 8.13 When $a = 1, b = 1$
  i) $\bar{a} = 0$
  ii) $a \wedge b = \text{Min}(1,1) = 1$
  iii) $a \vee b = \text{Max}(1,1) = 1$
  iv) $a \rightarrow b = \text{Min}(1, 1-1+1) = 1$ $\square$

Example 8.14 When a $= 0.6$, b $= 0.7$
  i) $\bar{a} = 0.4$
  ii) $a \wedge b = \text{Min}(0.6, 0.7) = 0.6$
  iii) $a \vee b = \text{Max}(0.6, 0.7) = 0.7$

iv) $a \rightarrow b = \text{Min}(1, 1-0.6+0.7) = \text{Min}(1, 1.1) = 1$ □

8.3 Linguistic Variable

8.3.1 Definition of Linguistic Variable

When we consider a variable, in general, it takes numbers as its value.   If the variable takes linguistic terms, it is called "linguistic variable".

Definition (Linguistic variable) The linguistic variable is defined by the following quintuple.
Linguistic variable = $(x, T(x), U, G, M)$
   x: name of variable
   $T(x)$: set of linguistic terms which can be a value of the variable
   U: set of universe of discourse which defines the characteristics of the variable
   G: syntactic grammar which produces terms in $T(x)$
   M: semantic rules which map terms in $T(x)$ to fuzzy sets in U   □

Example 8.15 Let's consider a linguistic variable "X" whose name is "Age".
   $X = (\text{Age}, T(\text{Age}), U, G, M)$
   Age: name of the variable X
   $T(\text{Age})$: {young, very young, very very young, …}
          Set of terms used in the discussion of age
   U: [0,100] universe of discourse
   $G(\text{Age})$: $T^{i+1} = \{\text{young}\} \cup \{\text{very } T^i\}$
   $M(\text{young}) = \{(u, \mu_{\text{young}}(u)) \mid u \in [0,100]\}$

$$\mu_{\text{young}}(u) = \begin{cases} 1 & \text{if } u \in [0,25] \\ (1 + \frac{u-25}{5})^{-2} & \text{if } u \in [25,100] \end{cases}$$ □

In the above example, the term "young" is used as a basis in the T(Age), and thus this kind of term is called a "primary term".   When we add modifiers to the primary terms, we can define new terms (fuzzy terms). In many cases, when the modifier "very" is added, the membership is obtained by square operation.   For example, the membership function of the term "very young" is obtained from that of "young".

$$\mu_{\text{very young}}(u) = (\mu_{\text{young}}(u))^2$$

The fuzzy linguistic terms often consist of two parts:

(1)   Fuzzy predicate(primary term): expensive, old, rare, dangerous, good, etc

(2) Fuzzy modifier: very, likely, almost impossible, extremely unlikely, etc

The modifier is used to change the meaning of predicate and it can be grouped into the following two classes:

(3) Fuzzy truth qualifier: quite true, very true, more or less true, mostly false, etc

(4) Fuzzy quantifier: many, few, almost, all, usually, etc

In the following sections, we will introduce the fuzzy predicate, fuzzy modifier, and fuzzy truth quantifier.

8.3.2 Fuzzy Predicate

As we know now, a predicate proposition in the classical logic has the following form.

"x is a man."

"y is P."

x and y are variables, and "man" and "P" are crisp sets.   The sets of individuals satisfying the predicates are written by "man(x)" and "P(y)".

Definition (Fuzzy predicate) .A fuzzy predicate is a predicate whose definition contains ambiguity   □

Example 8.16 For example,

"z is expensive."

"w is young."

The terms "expensive" and "young" are fuzzy terms.   Therefore the sets "expensive(z)" and "young(w)" are fuzzy sets.   □

When a fuzzy predicate "x is P" is given, we can interpret it in two ways.

(1) $P(x)$ is a fuzzy set.   The membership degree of x in the set P is defined by the membership function $\mu_{P(x)}$.

(2) $\mu_{P(x)}$ is the satisfactory degree of x for the property P.   Therefore, the truth value of the fuzzy predicate is defined by the membership function.

$$\text{Truth value} = \mu_{P(x)}$$

8.3.3 Fuzzy Modifier

As we know, a new term can be obtained when we add the modifier "very" to a primary term.   In this section we will see how semantic of the new term and membership function can be defined.

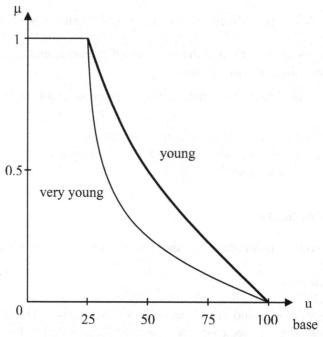

Fig. 8.1. Linguistic variable "Age"

Example 8.17 Let's consider a linguistic variable "Age" in (Fig 8.1.) Linguistic terms "young" and "very young" are defined in the universal set U.

$$U = \{u \mid u \in [0,100]\}$$

The variable Age takes a value in the set T(Age).

$$T(Age) = \{young, \text{ very young, very very young, } … \}$$

In the figure, the term "young" is represented by a membership function $\mu_{young}(u)$. When we represent the term "very young", we can use the square of $\mu_{young}(u)$ as follows.

$$\mu_{very\ young}(u) = (\mu_{young}(u))^2$$

The graph of membership function of "very young" is given in the figure.□

8.4 Fuzzy Truth Qualifier

8.4.1 Fuzzy Truth Values

Baldwin defined fuzzy truth qualifier in the universal set $V = \{v \mid v \in [0,1]\}$ as follows.

T = {true, very true, fairly true, absolutely true, ... , absolutely false, fairly false, false}

The qualifiers in T define "fuzzy truth values" and they can be defined by the membership functions. If we take baldwin's membership function $\mu_{true}(v)$, the truth qualifiers are represented by the followig membership functions(Fig 8.2).

$$\mu_{true}(v) = v \qquad\qquad v \in [0,1]$$
$$\mu_{very\ true}(v) = (\mu_{true}(v))^2 \qquad\qquad v \in [0,1]$$
$$\mu_{fairly\ true}(v) = (\mu_{true}(v))^{1/2} \qquad\qquad v \in [0,1]$$
$$\mu_{false}(v) = 1 - \mu_{true}(v) \qquad\qquad v \in [0,1]$$
$$\mu_{very\ false}(v) = (\mu_{false}(v))^2 \qquad\qquad v \in [0,1]$$
$$\mu_{fairly\ false}(v) = (\mu_{false}(v))^{1/2} \qquad\qquad v \in [0,1]$$
$$\mu_{absolutely\ true}(v) = \begin{cases} 1 & for\ v = 1 \\ 0 & otherwise \end{cases}$$
$$\mu_{absolutely\ false}(v) = \begin{cases} 1 & for\ v = 0 \\ 0 & otherwise \end{cases}$$

Example 8.18 Let's consider a predicate using the primary term "young" and fuzzy truth qualifier "very false"

$$P = \text{"Tom is young is very false."}$$

Suppose the term "young" is defined by the function μ_{young}.

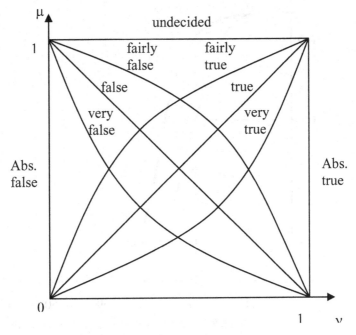

Fig. 8.2. Baldwin's truth graph

$$\mu_{\text{young}}(u) = \begin{cases} 1 & u \in [0,25] \\ (1+\frac{u-25}{5})^{-2} & u \in [25,100] \end{cases}$$

The term "very false" can be defined by the following.

$$\begin{aligned}\mu_{\text{very false}} &= (1 - \mu_{\text{true}}(u))^2 \\ &= (1 - \mu_{\text{young}}(u))^2 \\ &= \begin{cases} 0 & u \in [0,25] \\ (1-(1+\frac{u-25}{5})^{-2})^2 & u \in [25,100] \end{cases}\end{aligned}$$

Therefore, if Tom has age less than 25, the truth value of the predicate P is 0. If he is in [25,100], the truth value is calculated from $\mu_{\text{very false}}$.

8.4.2 Examples of Fuzzy Truth Qualifier

Example 8.19 Let's consider a predicate P in the following.

P = "20 is young."

Assume the terms "young" and "very young" are defined as shown in (Fig 8.3.)

We see the membership degree of 20 in "young" is 0.9. Therefore, the truth value of the predicate P is 0.9. Now we can modify the predicate P by using fuzzy truth qualifiers as follows.

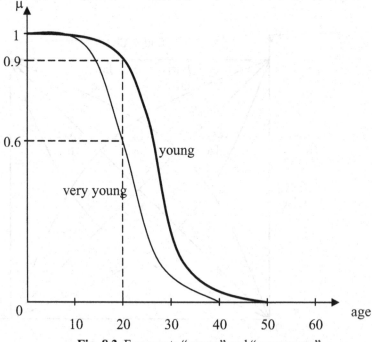

Fig. 8.3. Fuzzy sets "young" and "very young"

$$P_1 = \text{``20 is young is true.''}$$
$$P_2 = \text{``20 is young is fairly true.''}$$
$$P_3 = \text{``20 is young is very true.''}$$
$$P_4 = \text{``20 is young is false.''}$$

The truth values are changed according to the qualifiers as shown in (Fig 8.4.)

We know already $\mu_{young}(20)$ is 0.9. That is, the truth value of P is 0.9. For the predicate P_1, we use the membership function "true" in the figure and obtain the truth value 0.9. For P_2, the membership function "fairly true" is used and 0.95 is obtained. In the same way, we can calculate for P_3 and P_4 and summarize the truth values in the following.

For P_1: 0.9
For P_2: 0.95
For P_3: 0.81
For P_4: 0.1   □

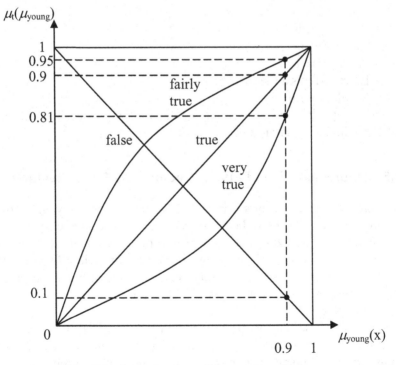

Fig. 8.4. The truth values of fuzzy proposition

8.5 Representation of Fuzzy Rule

8.5.1 Inference and Knowledge Representation

In general, the "inference" is a process to obtain new information by using existing knowledge. The representation of knowledge is an important issue in the inference. When we consider the representation methods, the following rule type "if-then" is the most popular form.

"If x is a, then y is b."

The rule is interpreted as an "implication" and consists of the "antecedent (if part)" and "consequent (then part)". If a rule is given in the above form and we have a fact in the following form,

"x is a"

then we can infer and obtain new result:

"y is b"

based on the above discussion, we can summarize two types of "reasoning".

(1)   Modus ponens
        Fact: x is a
        Rule: If x is a, then y is b
        Result: y is b
(2)   Modus tollens
        Fact: y is $\bar{b}$
        Rule: If x is a then y is b
        Result: x is $\bar{a}$

The modus ponens is used in the forward inference and the modus tollens is in the backward one.

8.5.2 Representation of Fuzzy Predicate by Fuzzy Relation

We saw that a fuzzy predicate is considered as a fuzzy set. In this section, we will see how the fuzzy predicate is used in fuzzy inference. When there is a fuzzy predicate proposition such that "x is P", it is represented by fuzzy set $P(x)$ and whose membership function is by $\mu_{P(x)}(x)$. We know also a fuzzy relation is one type of fuzzy sets, and thus we can represent a predicate by using a relation.

"$R(x) = P$"

P is a fuzzy set and $R(x)$ is a relation that consists of elements in P. The membership function of the predicate is represented by $\mu_{P(x)}(x)$ which shows the membership degree of x in P. The predicate represented by a relation will be used in the representation of fuzzy rule and premise.

8.5.3 Representation of Fuzzy Rule

When we consider fuzzy rules, the general form is given in the following.

If x is A, then y is B.

The fuzzy rule may include fuzzy predicates in the antecedent and consequent, and it can be rewritten as in the form.

If $A(x)$, then $B(y)$

This rule can be represented by a relation $R(x, y)$.

$R(x, y)$: If $A(x)$, then $B(y)$

or

$$R(x, y): A(x) \rightarrow B(y)$$

If there are a rule and facts involving fuzzy sets, we can execute two types of reasoning.

(1) Generalized modus ponens (GMP)

Fact:	x is A'	: $R(x)$
Rule:	If x is A then y is B	: $R(x, y)$
Result:	y is B'	: $R(y) = R(x) \circ R(x, y)$

(2) Generalized modus tollens (GMT)

Fact:	y is B'	: $R(y)$
Rule:	If x is A then y is B	: $R(x, y)$
Result:	x is A'	: $R(x) = R(y) \circ R(x, y)$

In the above reasoning, we see that the facts (A' and B') are not exactly same with the antecedents (A and B) in the rules; the results may be also different from the consequents. Therefore, we call this kind of inference as "fuzzy (approximate) reasoning or inference".

In general, when we execute the fuzzy (approximate) reasoning, we apply the "compositional rule of inference". The operation used in the reasoning is denoted by the notation "$\circ$", and thus the result is represented by the output of the composition when we use the GMP.

$$R(y) = R(x) \circ R(x, y)$$

Example 8.20   We have knowledge such as :

If x is A then y is B

x is A'

From the above knowledge, how can we apply the inference procedure to get new information about y ?

  i) We apply the implication operator to get implication relation

  $R(x,y) = A \times B$.   Here, the cartesian product A $\times$B is used.

  $R(x, y): A(x) \rightarrow B(y)$

  ii) We manipulate the fact into the form $R(x)$ and then apply the generalized modus ponens

$$R(y) = R(x) \circ R(x, y)$$

In this step, composition operator " $\circ$ " is used.   □

Therefore, there are two issues in the fuzzy reasoning: determination of the "implication relation" $R(x,y)$ and selection of the "composition operatior". These issues will be discussed in the next chapter.

[SUMMARY]

☐ Classical logic
  – proposition: a sentence having truth value true (1) or false (0)
  – truth value: proposition → {0, 1}
  – logic variable: variable representing a proposition

☐ Logic operation
  – negation (NOT): $\overline{P}$
  – conjunction (AND): $a \wedge b$
  – disjunction (OR): $a \vee b$
  – implication (→): $a \to b$

☐ Logic function
  – logic function
  – logic primitive
  – logic formula

☐ Tautology
  – tautology: logic formula whose value is always true
  – inference: developing new facts by using the tautology

☐ Deductive inference
  – modus ponens
  – modus tollens
  – hypothetical syllogism

☐ Predicate logic
  – predicate
  – predicate logic
  – predicate proposition: proposition consist of predicate and object
  – evaluation of proposition

☐ Quantifier
  – universal quantifier: $\forall$ (for all)
  – existential quantifier: $\exists$ (there exists)

☐ Fuzzy logic
  – fuzzy logic formula
  – fuzzy proposition

- truth value: fuzzy proposition $\rightarrow$ [0, 1]

☐ Fuzzy logic operation
  - negation (NOT): $\bar{P}$
  - conjunction (AND): $a \wedge b$
  - disjunction (OR): $a \vee b$
  - implication ($\rightarrow$): Min(1, 1+b−a)

☐ Linguistic variable
  - linguistic variable
  - linguistic terms (fuzzy sets)

☐ Fuzzy predicate
  - fuzzy predicate: predicate represented by fuzzy sets
  - fuzzy truth value [0, 1]
  - fuzzy modifier

☐ Fuzzy truth qualifier
  - fuzzy truth value: true, very true, fairly true, etc.
  - $\mu_{\text{very true}}(v) = (\text{true}(v))^2$
  - value of "P is very true" is 0.81 when value of P is 0.9

☐ Fuzzy rule
  - rule representation: if A(x) then B(y)
  - rule is a relation R(x, y)
  - inference: $R(y) = R(x) \circ R(x, y)$

☐ Fuzzy reasoning (inference)
  - approximate reasoning: facts may not equal to antecedents
  - generalized modus ponens
  - generalized modus tollens

[EXERCISES]

8.1 Evaluate the following propositions.
- a) $2 + 6 = 9$
- b) $3 + 4 = 7$
- c) $x + 5 = 7$
- d)   Einstein is a man.
- e)   The man has no head.

8.2 Develop the truth value of "exclusive OR".

8.3 Evaluate the following implication propositions.
- a) $a \rightarrow b$ where a : $2 + 2 = 5$,   b : $3 + 3 = 6$
- b) $a \rightarrow b$ where a : $3 + 4 = 6$,   b : $4 + 2 = 5$

8.4 Show truth values of the logic formula.
$$(\bar{b} \wedge (a \rightarrow b)) \rightarrow \bar{a}$$

8.5 Evaluate the following predicate propositions.
- a)   Sophie is a woman.
- b)   x is greater than 5 where x = 6.
- c)   y is green where y is a tree.
- d)   p belongs to me where p is a bag.

8.6 Evaluate the following fuzzy logic formulas where a = 0.5 and b = 0.7.
- a) $\bar{a} = 0.4$
- b) $a \wedge b$
- c) $a \vee b$
- d) $a \rightarrow b$

8.7 Define components for the linguistic variable X whose name is Temperature.
X = (Temperature, T(Temperature), U, G, M)

8.8 Determine the truth value of the following propositions P1 and P2.
P_1 = "P is very true"
P_2 = "P is false"
where
P = "30 is high",
the truth value of P is 0.3,
$\mu_{\text{very true}} = (\mu_{\text{true}})^2$

8.9 Show the following rule and fact in the form of relation and generalized modus ponens.

Rule: If $A(x)$ then $B(y)$: $R(x, y)$

Fact: x is A' : $R(x)$

Result: y is $B(y)$: $R(y) = R(x) \circ R(x, y)$

Chapter 9. FUZZY INFERENCE

In this chapter, we review the concepts of the extension principle and fuzzy relations, which expand the notions and applicability of fuzzy sets. By interpreting fuzzy rules as appropriate fuzzy relations, we investigate different schemes of fuzzy reasoning. The inference schemes are based on the compositional rule of inference, and the result is derived from a set of fuzzy rules and given inputs. The two important issues (determination of implication relation and selection of composition operator) will be discussed and four different inference methods will be introduced.

9.1 Composition of Rules

9.1.1 Extension Principle and Composition

As we studied in chap 3, the extension principle is a basic concept of fuzzy set theory that provides a general procedure for extending crisp domains of mathematical expressions to fuzzy domains. This procedure generalizes an ordinary mapping of a function f to a mapping between fuzzy sets. Suppose that g is a function from X to Y, and A is a fuzzy set on X defined as

$$A = \{(x_1, \mu_A(x_1)), (x_2, \mu_A(x_2)), \ldots, (x_n, \mu_A(x_n))\}$$

then the extension principle states that the image of fuzzy set A under the mapping f can be expressed as a fuzzy set $B \subseteq Y$.

$$B = f(A) = \{(y, \mu_B(y))\} \text{ where } \mu_B(y) = \max_{x = f^{-1}(y)} \mu_A(x)$$

We know that composition of sets is obtained by the Cartesian product of the sets. We can summarize various kinds of compositions discussed in chap 3.

(1) Composition of crisp sets A and B. It can represent a relation R between the sets A and B.

$$R = \{(x, y) \mid x \in A, y \in B\}, \quad R \subseteq A \times B$$

(2)   Composition of fuzzy sets A and B.   It is a relation R between A and
      B

$$R = \{((x, y), \mu_R(x, y)) \mid \mu_R(x, y) = \min[\mu_A(x), \mu_B(y)]$$
$$\text{or } \mu_R(x, y) = \mu_A(x) \cdot \mu_B(y)\}.$$

(3)   Composition of crisp relations R and S

$$S \circ R = \{(x, z) \mid (x, y) \in R, (y, z) \in S\}$$

where $R \subseteq A \times B$, $S \subseteq B \times C$, and $S \circ R \subseteq A \times C$.

(4)   Composition of fuzzy relations R and S

$$SR = S \circ R = \{((x, y), \mu_{SR}(x, z))\}$$

where $\mu_{SR}(x, z) = \max_{y} \min[\mu_R(x, y), \mu_S(y, z)]\}$.

The composition of fuzzy sets and relations will be elaborated in more
detail in the next sections.

9.1.2 Composition of Fuzzy Sets

Composition of fuzzy sets is obtained by Cartesian product of them.   For
the product space on fuzzy sets X and Y, there are two types of operations:
"fuzzy conjunction" and "fuzzy disjunction"

(1)   Fuzzy conjunction: the Cartesian product on X and Y is interpreted as
      a fuzzy conjunction defined by

$$A \times B = \int_{X \times Y} \mu_A(x) * \mu_B(y)/(x, y)$$

where $*$ is an operator representing a t-norm (triangle-norm), $x \in X$, $y \in$
Y, $A \subset X$, and $B \subset Y$.

(2)   Fuzzy disjunction: the Cartesian product on X × Y is interpreted as a
      disjunction defined by

$$A \times B = \int_{X \times Y} \mu_A(x) \dotplus \mu_B(y)/(x, y)$$

where $\dotplus$ is an operator representing a t-conorm (or s-norm), $x \in X$, $y \in$
Y, $A \subset X$, and $B \subset Y$.

The min and algebraic product operators are t-norm operators, and the
max operator is a t-conorm one.

9.1.3 Composition of Fuzzy Relations

As we know, unary fuzzy relations are fuzzy sets with one-dimensional
membership functions, and binary fuzzy relations are fuzzy sets with two-
dimensional membership functions.   Applications of the fuzzy relations

includes areas such as fuzzy control and decision making. A fuzzy relation $R \subseteq X \times Y$ defined in the Cartesian product space $X \times Y$ is determined by

$$R = \{((x, y), \mu_R(x, y)) \mid (x, y) \in X \times Y\}$$

Fuzzy relations in different product spaces can be combined through a "composition operation" denoted by the notation "$\circ$". Different composition operations have been suggested for fuzzy relations; we introduce two well known operations.

(1) The max-min composition is defined by

$$R_1 \circ R_2 = \{((x, z), \mu_{R1 \circ R2}(x, z))\}$$

where $\mu_{R1 \circ R2}(x, z) = \max_{y} \min [\mu_{R1}(x, y), \mu_{R2}(y, z)]$

$$= \bigvee_{y} [\mu_{R1}(x, y) \wedge \mu_{R2}(y, z)]$$

$$x \in X, y \in Y, z \in Z$$
$$R_1 \subseteq X \times Y, R_2 \subseteq Y \times Z$$

(2) The max-product composition is defined by

$$R_1 \circ R_2 = \{((x, z), \mu_{R1 \circ R2}(x, z))\}$$

where $\mu_{R1 \circ R2}(x, z) = \max_{y} [\mu_{R1}(x, y) \cdot \mu_{R2}(y, z)]$

$$x \in X, y \in Y, z \in Z$$
$$R_1 \subseteq X \times Y, R_2 \subseteq Y \times Z$$

The operator "$\cdot$" represents an algebraic product operation.

9.1.4 Example of Fuzzy Composition

Example 9.1 Let's consider a fuzzy rule in the following.
"x and y are approximately equal."
For this rule, a premise is given like
"x is small."
From the above facts, we want to know the knowledge about y.   First, we have to define the predicates included in the premise and rule. We know already relation R is a set and thus the premise can be represented by the form of relation $R(x)$. We assume that the variables x and y are positive integers in [1,4], and

$R(x, y) = \text{Approximately\_Equal}(x, y)$
$R(x) = \text{Small}(x)$

We also assume the membership degrees of $R(x)$ and $R(x, y)$ are given in (Tables 9.1, 9.2.)
For the fuzzy reasoning, let's use the max-min composition operator.

$$R(y) = R(x) \circ R(x, y)$$
$$\mu_R(y) = \bigvee_{x} [\mu_R(x) \wedge \mu_R(x, y)]$$

Table 9.1. Membership degrees of R(x)

x	1	2	3	4
$\mu_R(x)$	1	0.6	0.2	0

Table 9.2. Membership degrees of R(x, y)

x \ y	1	2	3	4
1	1	0.5	0	0
2	0.5	1	0.5	0
3	0	0.5	1	0.5
4	0	0	0.5	1

Through the composition operations, we obtain the result R(y) as shown in (Table 9.3.)   This reasoning procedure is called the fuzzy interence by for the generalized modus ponens.

That is,

> Generalized modus ponens (GMP)
>
> Input: R(x) in   (Table 9.1.)
>
> Rule: R(x, y) in (Table 9.2.)
>
> Result: R(y) in (Table 9.3.)   □

Table 9.3. Membership degrees of R(y)

y	1	2	3	4
$\mu_R(y)$	1	0.6	0.5	0.2

Example 9.2 Suppose the membership degrees of R(x) is given as shown in (Table 9.4.)   That is, the input is in the form of singleton value, $x = 2$. The inference result is given in (Table 9.5.) when input is given as $x = 2$ □

The result is obtained in the form of fuzzy set.   If there is a need to present the output as a form of linguistic term, we have to find a linguistic term which is the closest to the obtained fuzzy set.

Table9.4. Singleton input value $x = 2$

X	1	2	3	4
$\mu_R(x)$	0	1	0	0

Table 9.5. Inference result when $x = 2$

y	1	2	3	4
$\mu_R(y)$	0.5	1	0.5	0

To find the closest, we may use the measuring technique of distance between fuzzy sets, and this procedure is called the "linguistic approximation".   In the linguistic approximation, we try to find a linguistic term which has the minimum distance from the given fuzzy set.

9.2 Fuzzy Rules and Implication

9.2.1 Fuzzy if-then Rules

A fuzzy rule generally assumes the form
                    R: If x is A, then y is B.
where A and B are linguistic values defined by fuzzy sets on universe of discourse X and Y, respectively.   The rule is also called a "fuzzy implication" or fuzzy conditional statement.   The part "x is A" is called the "antecedent" or "premise", while "y is B" is called the "consequence" or "conclusion".   In general, the antecedent and consequence are represented by the form of linguistic variables discussed in the previous chapter.

   Before we employ fuzzy if-then rules to model and analyze a system, first we have to formalize what is meant by the expression
                    R: "If x is A then y is B",
which is sometimes abbreviated as
                    R: A $\rightarrow$ B
   In essence, the expression describes a relation between two variables x and y.   This suggests that a fuzzy rule can be defined as a binary relation R on the product space X $\times$ Y.

9.2.2 Fuzzy Implications

Based on the interpretations of the Cartesian product and various t-norm and t-conorm operators, a number of qualified methods can be formulated to calculate the fuzzy relation

$$R = A \rightarrow B$$

R can be viewed as a fuzzy set with a two-dimensional membership function

$$\mu_R(x, y) = f(\mu_A(x), \mu_B(y))$$

where the function f, called the "fuzzy implication function", performs the task of transforming the membership degrees of x in A and y in B into those of (x, y) in A × B. We introduce here two well known fuzzy implication functions.

(1) Min operation rule of fuzzy implication [Mamdani]. It interprets the fuzzy implication as the minimum operation.

$$R_C = A \times B$$

$$= \int_{X \times Y} \mu_A(x) \wedge \mu_B(y) / (x, y)$$

where $\wedge$ is the min operator

(2) Product operation rule of fuzzy implication [Larsen]. It implements the implication by the product operation.

$$R_P = A \times B$$

$$= \int_{X \times Y} \mu_A(x) \cdot \mu_B(y) / (x, y)$$

where $\cdot$ is the algebraic product operator

When we use the conjunction for the Cartesian product on X and Y, Mamdani's min fuzzy implication R_C is obtained if the minimum operator is used for the conjunction; Larsen's product fuzzy implication R_P is obtained if the algebraic product is used. The two implication functions R_C and R_P are often adopted functions in the fuzzy reasoning.

9.2.3 Example of Fuzzy Implications

Example 9.3 There is a fuzzy rule in the following.
            If temperature is high, then humidity is fairly high.

It is a fuzzy rule and a fuzzy relation. We want to determine the membership function of the rule. Let T and H be universe of discourse of temperature and humidity, respectively, and let's define variables $t \in T$ and $h \in H$. We represent the fuzzy terms "high" and "fairly high" by A and B respectively:

$$A = \text{"high"}, \qquad A \subseteq T$$

$$B = \text{"fairly high"}, \quad B \subseteq H$$

Table 9.6. Membership of A in T (temperature)

t	20	30	40
$\mu_A(t)$	0.1	0.5	0.9

Table 9.7. Membership degrees of B in H (humidity)

h	20	50	70	90
$\mu_B(h)$	0.2	0.6	0.7	1

then the above rule can be rewritten as

$$R(t, h): \text{If t is A, then h is B.}$$

In the rule (relation), we can find two predicate propositions:

$$R(t): \text{t is A}$$
$$R(h): \text{h is B}$$

the rule becomes

$$R(t, h): R(t) \rightarrow R(h)$$

if we know membership functions of A and B, we can determine $R_{(t, h)} = A \times B$ by using the fuzzy implication function where $R_{(t, h)} \subseteq T \times H$. Assume membership functions $\mu_A(t)$ and $\mu_B(h)$ are given in (Tables 9.6 , 9.7) respectively.

In order to get the relation for the implication in the above fuzzy rule, we have to select an implication function between A and B. For simplicity, let's take the min operation of Mamdani in the previous section.

$$R_C(t, h) = A \times B$$
$$= \int \mu_A(t) \wedge \mu_B(h) / (t, h)$$

when we apply the min operation on the Cartesian product $A \times B$, we obtain the relation R_C as shown in (Table 9.8.) This membership of R_C represents the fuzzy rule. Note that $\mu_{R_C}(20, 50) = 0.1$ is obtained by the min between $\mu_A(20) = 0.1$ and $\mu_B(50) = 0.6$. Similarly, $\mu_{R_C}(30, 20) = 0.2$ from $\mu_A(30) = 0.5$ and $\mu_B(20) = 0.2$. □

Example 9.4 Now suppose, we want to get information about the humidity when there is the following premise about the temperature.

$$\text{"Temperature is fairly high"}$$

This fact is rewritten as

R(t): "t is A'"   where A' = "fairly high"
where the fuzzy term A' ⊆ T is defined in (Table 9.9)

Table 9.8. Membership of rule $R_C = A \times B$

t \ h	20	50	70	90
20	0.1	0.1	0.1	0.1
30	0.2	0.5	0.5	0.5
40	0.2	0.6	0.7	0.9

Table 9.9. Membership function of A' in T (temperature)

t	20	30	40
$\mu_{A'}(t)$	0.01	0.25	0.81

As we can see, A' is not same with A and thus we apply the fuzzy inference method of generalized modus ponens. We use the composition rule of inference with the max-min composition.

R(h) = R(t) ∘ R_C(t, h)

where R(t) is in (Table 9.9.) and R_C(t, h) in (Table 9.8.)

If we denote the result of the inference as B', B' is the information about humidity when "temperature is fairly high" (Table 9.10).   □

9.3 Inference Mechanism

9.3.1 Decomposition of Rule Base

When we model a knowledge system, it is often represented by the form of "fuzzy rule base". The fuzzy rule base consists of fuzzy if-then rules. In many cases, the fuzzy reasoning on the fuzzy rule base is based on one-level forward data-driven inference (GNP: generalized modus ponens).

The rule base has the form of a MIMO (multiple input multiple output) system.

Table 9.10. Result of fuzzy inference

h	20	50	70	90
$\mu_{B'}(h)$	0.2	0.6	0.7	0.81

$$R = \{R^1_{MIMO}, R^2_{MIMO}, \ldots, R^n_{MIMO}\}$$

where R^i_{MIMO} represents the rule:

If x is A_i and y is B_i, then z_1 is C_i, ..., z_q is D_i,

The antecedent of R^i_{MIMO} forms a fuzzy set $A_i \times \ldots \times B_i$ in the "product space" $U \times \ldots \times V$. The consequence is the "union" of q independent control actions $(z_1 + z_2 + \ldots + z_q)$. Thus the ith rule R^i_{MIMO} may be represented as a fuzzy implication.

$$R^i_{MIMO} : (A_i \times \ldots \times B_i) \rightarrow (z_1 + \ldots + z_q)$$

From the above statement, it follows that the rule base R may be represented as the union

$$R = \{\bigcup_{i=1}^{n} R^i_{MIMO}\}$$

$$= \{\bigcup_{i=1}^{n} [(A_i \times \cdots \times B_i) \rightarrow (z_1 + \cdots + z_q)]\}$$

$$= \{\bigcup_{i=1}^{n} [(A_i \times \cdots \times B_i) \rightarrow z_1],$$

$$\bigcup_{i=1}^{n} [(A_i \times \cdots \times B_i) \rightarrow z_2], \ldots,$$

$$\bigcup_{i=1}^{n} [(A_i \times \cdots \times B_i) \rightarrow z_q]\}$$

$$= \{\bigcup_{k=1}^{q} \bigcup_{i=1}^{n} [(A_i \times \cdots \times B_i) \rightarrow z_k]\}$$

$$= \{\bigcup_{k=1}^{q} RB^k_{MISO}\} \quad where \; RB^k_{MISO} = \bigcup_{i=1}^{n} [(A_i \times \cdots \times B_i) \rightarrow z_i]$$

$$= \{RB^1_{MISO}, RB^2_{MISO}, \cdots, RB^k_{MISO}, \cdots, RB^q_{MISO}\}$$

In effect, the rule base R is composed of a set of sub-rule-bases RB^k_{MISO} where k = 1, 2, ... , q. The sub-rule-base RB^k_{MISO} has "multiple input" variables and a "single control" variable. Therefore the general rule structure of a MIMO fuzzy system can be represented as a collection of MISO fuzzy systems.

$$R = \{RB^1_{MISO}, RB^2_{MISO}, \cdots, RB^k_{MISO}, \cdots, RB^q_{MISO}\}$$

where RB^k_{MISO} represents the rule:

If x is A_i and ... , and y is B_i then z_k is C_i,   for i = 1, 2, ... , n

9.3.2 Two-Input/Single-Output Rule Base

For simplicity, let's consider the general form of MISO fuzzy control rules in the case of two-input/single-output systems.

Input: u is A' and v is B'

R_1: if u is A_1 and v is B_1 then is w is C_1

else R_2: if u is A_2 and v is B_2 then is w is C_2

•••

•••

else R_n: if u is A_n and v is B_n then is w is C_n

consequence: w is C'

where u, v, and w are linguistic variables representing the process state variables and the control variables, respectively. A_i, B_i, and C_i are linguistic values of the linguistic variables u, v, and w in the universe of discourse U, V, and W respectively for i=1, 2, ... , n.

The fuzzy control rule

R_i: If u is A_i and v is B_i then w is C_i

is implemented as a fuzzy implication relation R_i and is defined as

R_i: $(A_i$ and $B_i) \rightarrow C_i$ or

$$\mu_{R_i} = \mu_{(A_i \text{ and } B_i \rightarrow C_i)}(u, v, w)$$

$$= [\mu_{A_i}(u) \text{ and } \mu_{B_i}(v)] \rightarrow \mu_{C_i}(w)$$

where "A_i and B_i" is a fuzzy set $A_i \times B_i$ in U × V.

R_i: $(A_i$ and $B_i) \rightarrow C_i$ is a fuzzy implication relation in U × V × W, and $\rightarrow$ denotes a fuzzy implication function.

9.3.3 Compositional Rule of Inference

Let's consider a single fuzzy rule and its inference (GMP: generalized modus ponens).

R_1: if v is A then w is C

Input: v is A'

Result: C'

$A \subset U, C \subset W, v \in U$, and $w \in C$. The fuzzy rule is interpreted as an implication $(A \rightarrow C)$ and which is defined in the product space U × W.

$R_1 : A \rightarrow C$ or $R_1 = A \times C$

$R_1 \subset U \times W$

When the input A' is given to the inference system, the output C' is obtained through the inference operation denoted by the composition operator "∘".

$C' = A' \circ R_1$

This inference procedure is called as the "compositional rule of inference". Therefore in real systems, the inference is determined by two factors: the "implication operator" and "composition operator". As we saw in the previous section, for the implication in the Cartesian product space $R = A \times C$, the two operators are often used:

Mamdani implication (R_C): min operator

Larsen implication (R_P): algebraic product operator

for the composition, we have introduced also two operators.

Mamdani composition: max-min

Larsen composition: max-product

The following lemmas will show the inference mechanism when a simple fuzzy rule is given.

Lemma 1 (For 1 singleton input, result C' is obtained from C and matching degree α_1, Fig 9.1)

When a fuzzy rule R_1 and singleton input u_0 are given

R_1: If u is A then w is C,

Or R_1: $A \rightarrow C$

The inference result C' is defined by the membership function $\mu_C(w)$

$\mu_C(w) = \alpha_1 \wedge \mu_C(w)$ for R_C (Mamdani implication)

$\mu_C(w) = \alpha_1 \cdot \mu_C(w)$ for R_P (Larsen implication)

where $\alpha_1 = \mu_A(u_0)$

(Proof) $\mu_{R_1}(u, w) = \mu_A(u) \rightarrow \mu_C(w)$

By the compositional rule of inference,

$$C' = A' \circ (A \rightarrow C) = A' \circ R_1$$

In this case, $A' = u_0$

$$\mu_C(w) = \mu_0 \circ (\mu_A(u) \rightarrow \mu_C(w))$$
$$= \mu_A(u_0) \rightarrow \mu_C(w)$$

i) If we apply min operator for the implication,

$$\mu_{C'}(w) = \min[\mu_A(u_0), \mu_C(w)]$$
$$= \alpha_1 \wedge \mu_C(w) \quad where \ \alpha_1 = \mu_A(u_0)$$

ii) If we use Larsen's product operator for the implication,

$$\mu_{C'}(w) = \mu_A(u_0) \rightarrow \mu_C(w)$$
$$= \mu_A(u_0) \cdot \mu_C(w)$$
$$= \alpha_1 \cdot \mu_C(w) \quad where \ \alpha_1 = u_A(u_0)$$

The α_1 is called the "matching degree", "satisfaction degree", or "firing strength". □

Lemma 2: (For 1 fuzzy input, result C' is obtained from C and matching degree α_1, Fig 9.2)

When a fuzzy rule R_1: $A \rightarrow C$ and input A' are given, the inference result C' is defined by the membership function $\mu_{C'}$.

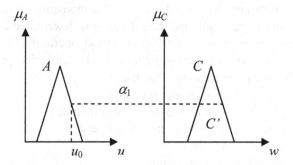

Fig. 9.1. Graphical representation of Lemma 1 with R_C
(When a singleton input is given, C' is obtained from C and α_1)

$$\mu_{C'}(w) = \alpha_1 \wedge \mu_C(w) \quad \text{for } R_C$$
$$\mu_{C'}(w) = \alpha_1 \cdot \mu_C(w) \quad \text{for } R_P$$
$$\text{where } \alpha_1 = \max_u [\mu_{A'}(u) \wedge \mu_A(u)]$$

(Proof) For R_C, we apply the minimum operator for the implication and the max-min for the composition.

$$C' = A' \circ (A \to C)$$
$$= A' \circ (A \times C)$$
$$= A' \circ R_1$$
$$\mu_{R_1}(u, w) = \min[\mu_A(u), \mu_C(w)] = [\mu_A(u) \wedge \mu_C(w)]$$
$$\mu_{C'}(w) = \{\mu_{A'}(u) \circ (\mu_A(u) \to \mu_C(w))\} = \{\mu_{A'}(u) \circ \mu_{R_1}(u, w)\}$$
$$= \max_u \{\mu_{A'}(u) \wedge \mu_{R_1}(u, w)\}$$
$$= \max_u \{\mu_{A'}(u) \wedge [\mu_A(u) \wedge \mu_C(w)]\}$$

As we know, the max and min operators are associative and thus

$$\mu_{C'}(w) = \max_u \{[\mu_{A'}(u) \wedge \mu_A(u)] \wedge \mu_C(w)\}$$
$$= \max_u [\mu_{A'}(u) \wedge \mu_A(u)] \wedge \mu_C(w)$$
$$= \alpha_1 \wedge \mu_C(w) \quad where \ \alpha_1 = \max_u [\mu_{A'}(u) \wedge \mu_A(u)]$$

Similarly, we can apply the Larsen product R_P and obtain the above result. □

9.3.4 Fuzzy Inference with Rule Base

In this section, we generalize the properties of compositional rule of inference discussed in the previous section to the case such as:

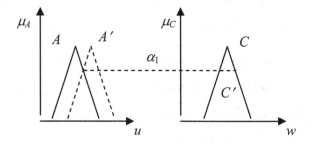

Fig. 9.2. Graphical representation of Lemma 2 with R_C
(When a fuzzy set input is given, C' is obtained from C and α_1)

$$R: \overset{n}{\underset{i=1}{U}} R_i$$

$$R_i: A_i \rightarrow C_i$$

Note that there are connectives "also" or "else" between the rules. Let's see how the connectives are interpreted when we consider the max-min for the composition operator.

We will see that the operators are commutative and thus the fuzzy control action is considered as the aggregated result derived from individual control rules.   That is, the connectives are implemented as max operations.   Furthermore, we shall also see the same properties even with the max-product operator for the composition.

Lemma 3 (Total result C' is an aggregation of individual results C_i', Fig 9.3)

The result of inference C is an aggregation of result C_i' derived from individual rules.

$$C' = A' \circ \overset{n}{\underset{i=1}{U}} R_i = \overset{n}{\underset{i=1}{U}} A' \circ R_i = \overset{n}{\underset{i=1}{U}} C_i'$$

(Proof)

$$C' = A' \circ \overset{n}{\underset{i=1}{U}} R_i = A' \circ \overset{n}{\underset{i=1}{U}} (A_i \rightarrow C_i)$$

$$R_i = A_i \rightarrow C_i \quad or \quad R_i = A_i \times C_i \quad for\ i = 1, 2, \cdots, n$$

The membership function $\mu_{C'}$ of the fuzzy set C' is pointwise defined for all $w \in W$ by

$$\mu_{C'}(w) = \mu_{A'}(u) \circ \max_{u, w}[\mu_{R_1}(u, w), \mu_{R_2}(u, w), \cdots, \mu_{R_n}(u, w)]$$

we replace the $\circ$ operator by max-min operator, and then

$$\mu_{C'}(w) = \max_u \min_u \{\mu_{A'}(u), \max_{u,w}[\mu_{R_1}(u,w), \mu_{R_2}(u,w), \cdots, \mu_{R_n}(u,w)]\}$$

$$= \max_u \max_{u,w}\{\min_u[\mu_{A'}(u), \mu_{R_1}(u,w)],$$

$$\min[\mu_{A'}(u), \mu_{R_2}(u,w)], \cdots,$$

$$\min[\mu_{A'}(u), \mu_{R_n}(u,w)]\}$$

Because $[\mu_{A'}(u) \circ \mu_{R_i}(u,w)] = \max_u \min_u[(\mu_{A'}(u), \mu_{R_i}(u,w)],$

$$\mu_{C'}(w) = \max_{u,w}\{[\mu_{A'}(u) \circ \mu_{R_1}(u,w)],$$

$$[\mu_{A'}(u) \circ \mu_{R_2}(u,w)], \cdots,$$

$$[\mu_{A'}(u) \circ \mu_{R_n}(u,w)]\}$$

Because

$$\mu_{C_i'} = [\mu_{A'}(u) \circ \mu_{R_i}(u,w)]$$

$$C_i' = A' \circ R_i$$

then

$$C' = [A' \circ R_1] \cup [A' \circ R_2] \cup \cdots \cup [A' \circ R_n]$$

$$= \bigcup_{i=1}^{n} A' \circ R_i$$

$$= \bigcup_{i=1}^{n} A' \circ (A_i \to C_i)$$

$$= \bigcup_{i=1}^{n} C_i'$$

We saw that the result C' is an union (aggregation) of result C_i' from individual rules.   That is,

$$\mu_{C'}(w) = \bigvee_{i=1}^{n} \mu_{C_i'} \qquad \Box$$

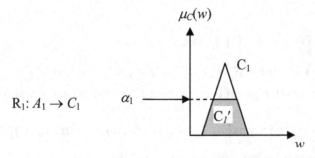

Fig. 9.3. Lemma 3 (Total result C' is a union of individual result C_i')

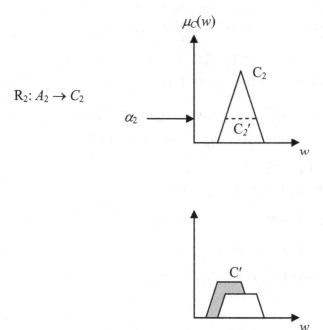

$R_2: A_2 \rightarrow C_2$

Fig. 9.3. (cont')

Now, we generalize Lemma 3 to the case of multiple input variables such as

$$R: \overset{n}{\underset{i=1}{U}} R_i$$

$R_i: A_i$ and $B_i \rightarrow C_i$

Colorally of Lemma 3: (Lemma 3 in the case of multiple inputs)
The result of inference C is an aggregation of result C_i' derived from individual rules.

$$C' = (A', B') \circ \overset{n}{\underset{i=1}{U}} R_i = \overset{n}{\underset{i=1}{U}} (A', B') \circ R_i = \overset{n}{\underset{i=1}{U}} C_i'$$

(Proof)

$$C' = (A', B') \circ \overset{n}{\underset{i=1}{U}} R_i = (A', B') \circ \overset{n}{\underset{i=1}{U}} (A_i \ and \ B_i \rightarrow C_i)$$

$$R_i = A_i \ and \ B_i \rightarrow C_i \quad or \quad R_i = A_i \times B_i \times C_i \quad for \ i = 1, 2, \cdots, n$$

The membership function $\mu_{C'}$ of the fuzzy set C' is pointwise defined for all $w \in W$ by

$$\mu_{C'}(w) = [\mu_{A'}(u), \mu_{B'}(v)] \circ \underset{u,v,w}{\max}[\mu_{R_1}(u,v,w), \mu_{R_2}(u,v,w), \cdots, \mu_{R_n}(u,v,w)]$$

we replace the $\circ$ operator by max-min operator, and then

$$\mu_{C'}(w) = \max_{u,v} \min_{u,v}\{[\mu_{A'}(u), \mu_{B'}(v)], \max_{u,v,w}[\mu_{R_1}(u,v,w),$$

$$\mu_{R_2}(u,v,w), \cdots, \mu_{R_n}(u,v,w)]\}$$

$$= \max_{u,v} \max_{u,v,w}\{\min_{u,v}[(\mu_{A'}(u), \mu_{B'}(v)), \mu_{R_1}(u,v,w)],$$

$$\min[(\mu_{A'}(u), \mu_{B'}(v)), \mu_{R_2}(u,v,w)], \cdots,$$

$$\min[(\mu_{A'}(u), \mu_{B'}(v)), \mu_{R_n}(u,v,w)]\}$$

Because $[(\mu_{A'}(u), \mu_{B'}(v)) \circ \mu_{R_i}(u,v,w)]$

$$= \max_{u,v} \min_{u,v}[(\mu_{A'}(u), \mu_{B'}(v), \mu_{R_i}(u, v, w)],$$

$$\mu_{C'}(w) = \max_{u,v,w}\{[(\mu_{A'}(u), \mu_{B'}(v)) \circ \mu_{R_1}(u,v,w)],$$

$$[(\mu_{A'}(u), \mu_{B'}(v)) \circ \mu_{R_2}(u,v,w)], \cdots,$$

$$[(\mu_{A'}(u), \mu_{B'}(v)) \circ \mu_{R_n}(u,v,w)]\}$$

Because

$$\mu_{C_i'} = [(\mu_{A'}(u), \mu_{B'}(v)) \circ \mu_{R_i}(u, v, w)]$$

$$C_i' = (A', B') \circ R_i$$

then

$$C' = [(A', B') \circ R_1] \cup [(A', B') \circ R_2] \cup \cdots \cup [(A', B') \circ R_n]$$

$$= \overset{n}{\underset{i=1}{U}}(A', B') \circ R_i$$

$$= \overset{n}{\underset{i=1}{U}}(A', B') \circ (A_i \ and \ B_i \to C_i)$$

$$= \overset{n}{\underset{i=1}{U}} C_i'$$

that is,

$$\mu_{C'}(w) = \overset{n}{\underset{i=1}{V}} \mu_{C_i'} \qquad \square$$

Lemma 4: $(R_i: (A_i \times B_i \to C_i)$ consists of $R_i^1: (A_i \to C_i)$ and $R_i^2: (B_i \to C_i)$, Fig 9.4)

When there is a rule R_i with two inputs variables A_i and B_i, the inference result C_i' is obtained from individual inferences of $R_i^1: (A_i \to C_i)$ and $R_i^2: (B_i \to C_i)$.

$$C_i' = (A', B') \circ (A_i \text{ and } B_i \to C_i)$$

$$= [A' \circ (A_i \to C_i)] \cap [B' \circ (B_i \to C_i)] \qquad \text{if } \mu_{A_i \times B_i} = \mu_{A_i} \wedge \mu_{B_i} \quad (\text{for } R_C)$$

$$= [A' \circ R_i^1] \cap [B' \circ R_i^2] \qquad \text{where } R_i^1 = A_i \to C_i \text{ and } R_i^2 = B_i \to C_i$$

$$= C_i^1 \cap C_i^2 \qquad \text{where } C_i^1 = A' \circ R_i^1 \text{ and } C_i^2 = A' \circ R_i^2$$

$$C_i' = (A', B') \circ (A_i \text{ and } B_i \to C_i)$$

$$= [A' \circ (A_i \to C_i)] \cdot [B' \circ (B_i \to C_i)] \qquad \text{if } \mu_{A_i \times B_i} \doteq \mu_{A_i} \cdot \mu_{B_i} \quad (\text{for } R_P)$$

(Proof) We prove the lemma in the case of R_C.

$$C_i' = (A', B') \circ (A_i \text{ and } B_i \to C_i)$$

$$\mu_{C_i'} = (\mu_{A'}, \mu_{B'}) \circ (\mu_{A_i \times B_i} \to \mu_C)$$

$$= (\mu_{A'}, \mu_{B'}) \circ (\min(\mu_{A_i}, \mu_{B_i}) \to \mu_{C_i})$$

$$= (\mu_{A'}, \mu_{B'}) \circ \min[(\mu_{A_i} \to \mu_{C_i}), (\mu_{B_i} \to \mu_{C_i})]$$

If we replace $\circ$ operator by the max-min operator,

$$\mu_{C_i'} = \max_{u,v} \min\{(\mu_{A'}, \mu_{B'}), \min[(\mu_{A_i} \to \mu_{C_i}), (\mu_{B_i} \to \mu_{C_i})]\}$$

$$= \max_{u,v} \min\{\min[\mu_{A'}, (\mu_{A_i} \to \mu_{C_i})], \min[\mu_{B'}, (\mu_{B_i} \to \mu_{C_i})]\}$$

$$= \min\{[\mu_{A'} \circ (\mu_{A_i} \to \mu_{C_i})], [\mu_{B'} \circ (\mu_{B_i} \to \mu_{C_i})]\}$$

Hence we have

$$C_i' = [A' \circ (A_i \to C_i)] \cap [B' \circ (B_i \to C_i)]$$

$$= [A' \circ R_i^1] \cap [B' \circ R_i^2]$$

$$= C_i^1 \cap C_i^2$$

This means that the inference result Ci′ is obtained by the intersection between individual inferences

A′ ∘ (Aᵢ → Cᵢ) and B′ ∘ (Bᵢ → Cᵢ).   Therefore, in the next lemma, we will show that we can apply min operator between $\mu_{A_i}(u_0)$ and $\mu_{B_i}(v_0)$ where there are two singleton inputs u_0 and v_0.   □

Lemma 5: (For singleton input, C_i' is determined by the minimum matching degree of Aᵢ and Bᵢ, Fig 9.5)

If the inputs are fuzzy singletons, namely, A′ = u0, B′ = v0, the matching degree αi is the minimum value between .$\mu_{Ai}(u_0)$ and $\mu_{Bi}(v_0)$ from the lemma 1, the inference result can be derived by employing Mamdani's minimum operation rule R_C and Larsen's product operation rule R_P for the implication.

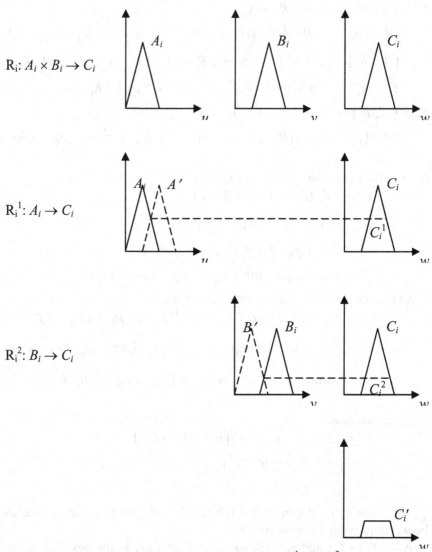

$R_i: A_i \times B_i \to C_i$

$R_i^1: A_i \to C_i$

$R_i^2: B_i \to C_i$

Fig. 9.4. Lemma 4 (Rule R_i can be decomposed into R_i^1 and R_i^2 and the result C_i' of R_i is an intersection of the results C_i^1 and C_i^2 of R_i^1 and R_i^2, respectively.)

$$\mu_{C_i'}(w) = \alpha_i \wedge \mu_{C_i}(w) \quad for\ R_C$$
$$\mu_{C_i'}(w) = \alpha_i \bullet \mu_{C_i}(w) \quad for\ R_P$$
$$where\ \alpha_i = \mu_{A_i}(u_0) \wedge \mu_{B_i}(v_0) = \min[\mu_{A_i}(u_0), \mu_{B_i}(v_0)]$$

(proof) From Lemma 4,

$$C_i' = [A' \circ (A_i \to C_i)] \cap [B' \circ (B_i \to C_i)]$$

$$\mu_{C_i'} = \min\{[u_0 \circ (\mu_{A_i}(u) \to \mu_{C_i}(w))], [v_0 \circ (\mu_{B_i}(v) \to \mu_{C_i}(w))]\}$$

$$= \min\{[\mu_{A_i}(u_0) \to \mu_{C_i}(w)], [\mu_{B_i}(v_0) \to \mu_{C_i}(w)]\}$$

1. If we use the Mamdani's minimum operation for the implication,

$$\mu_{C_i'} = \min\{\min[\mu_{A_i}(u_0), \mu_{C_i}(w)], \min[\mu_{B_i}(v_0), \mu_{C_i}(w)]\}$$

$$= \min\{\min[\mu_{A_i}(u_0), \mu_{B_i}(v_0), \mu_{C_i}(w)]\}$$

$$= \min\{\min[\mu_{A_i}(u_0), \mu_{B_i}(v_0)], \mu_{C_i}(w)\}$$

$$Because \ \alpha_i = \min[\mu_{A_i}(u_0), \mu_{B_i}(v_0)],$$

$$\mu_{C_i'} = \min\{\alpha_i, \mu_{C_i}(w)\}$$

$$= \alpha_i \wedge \mu_{C_i}(w)$$

2. If we use the Larsen's product operator for the implication,

$$\mu_{C_i'} = \min\{[\mu_{A_i}(u_0) \bullet \mu_{C_i}(w)], [\mu_{B_i}(v_0) \bullet \mu_{C_i}(w)]\}$$

$$= \{\min[\mu_{A_i}(u_0), \mu_{B_i}(v_0)]\} \bullet \mu_{C_i}(w)$$

$$= \alpha_i \bullet \mu_{C_i}(w)$$

Lemma 6: (For fuzzy input, C_i' is determined by the minimum matching degree of (A' and A_i) and (B' and B_i), Fig 9.6)

If the inputs are given as fuzzy sets A' and B', the matching degree α_i is determined by the minimum between (A' and A_i) and (B' and B_i). From the lemma 2, the results can be derived by employing the min operation for R_C and the product operation for R_P.

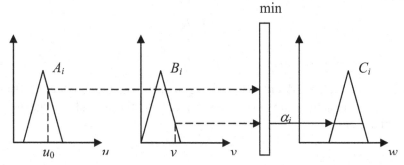

Fig. 9.5. Lemma 5 (α_i is the minimum matching degree between $A_i(u_0)$ and $B_i(v_0)$.)

$$\mu_{C_i'}(w) = \alpha_i \wedge \mu_{C_i}(w) \qquad for\ R_C$$

$$\mu_{C_i'}(w) = \alpha_i \cdot \mu_{C_i}(w) \qquad for\ R_P$$

$$where\ \alpha_i = \min[\max_u(\mu_{A'}(u) \wedge \mu_{A_i}(u)), \max_v(\mu_{B'}(v) \wedge \mu_{B_i}(v))]$$

(proof) From Lemma 5,

$$C_i' = [A' \circ (A_i \rightarrow C_i)] \cap [B' \circ (B_i \rightarrow C_i)]$$

$$\mu_{C_i'} = \min\{[\mu_{A'} \circ (\mu_{A_i} \rightarrow \mu_{C_i})], [\mu_{B'} \circ (\mu_{B_i} \rightarrow \mu_{C_i})]\}$$

$$= \min\{(\mu_{A_i} \circ \mu_{R_{AC}}), (\mu_{B_i} \circ \mu_{R_{BC}})\}$$

$$where\ R_{AC} = A_i \times C_i,\ R_{BC} = B_i \times C_i$$

If we apply the min for the implication (Cartesian product for R_{AC} and R_{BC}) and the max-min for the composition, from Lemma 2, we have

$$\mu_{C_i'} = \min\{\max_u \mu_{A_i'} \wedge \mu_{R_{AC}}, \max_v \mu_{B_i'} \wedge \mu_{R_{BC}}\}$$

$$= \min\{\max_u \mu_{A_i'} \wedge (\mu_{A_i} \wedge \mu_{C_i}), \max_v \mu_{B_i'} \wedge (\mu_{B_i} \wedge \mu_{C_i})\}$$

$$= \min\{\max_u(\mu_{A_i'} \wedge \mu_{A_i}) \wedge \mu_{C_i}, \max_v(\mu_{B_i'} \wedge \mu_{B_i}) \wedge \mu_{C_i}\}$$

$$= \min\{\alpha_A \wedge \mu_{C_i}, \alpha_B \wedge \mu_{C_i}\}$$

$$where\ \alpha_A = \max_u(\mu_{A_i'} \wedge \mu_{A_i}),\ \alpha_B = \max_v(\mu_{B_i'} \wedge \mu_{B_i})$$

$$= \min[\alpha_A, \alpha_B] \wedge \mu_{C_i} = \alpha_i \wedge \mu_{C_i}$$

$$where\ \alpha_i = \min[\alpha_A, \alpha_B] = \min[\max_u(\mu_{A_i'} \wedge \mu_{A_i}), \max_v(\mu_{B_i'} \wedge \mu_{B_i})]$$

Similarly, we can prove the lemma when the product is used for the implication and the max-product for the composition (R_P).
Therefore, from the lemmas, we can state

$$\mu_{C'} = \overset{n}{\underset{i=1}{U}} \alpha_i \wedge \mu_{C_i} \qquad for\ R_C$$

$$\mu_{C'} = \overset{n}{\underset{i=1}{U}} \alpha_i \cdot \mu_{C_i} \qquad for\ R_P$$

where the matching degree (weighting factor, satisfaction factor, firing strength) α_i is a measure of the contribution of the ith rule to the fuzzy control action.   □

9.4 Inference Methods

Based upon the previous lemmas, now we develop inference methods.

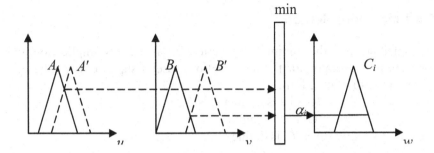

Fig. 9.6. Lemma 6 (α_i is the minimum matching degree between (A' and A_i) and (B' and B_i).)

For example, we consider two fuzzy control rules.

R_1: if u is A_1 and v is B_1 then w is C_1

R_2: if u is A_2 and v is B_2 then w is C_2

A_i, B_i, and C_i are defined in U,V, and W, respectively, for i = 1, 2, $u \in$ U, $v \in$ V, and $w \in$ W.

The inputs are usually measured by sensors and are crisp. In some cases it may be expedient to convert the input data into fuzzy sets. In general, however, a crisp value may be treated as a fuzzy singleton.

(1)   Singleton input

If the input are given as singleton values, the matching degrees (firing strengths) α_1 and α_2 of the first and second rules may be expressed as

$$\alpha_1 = \mu_{A_1}(u_0) \wedge \mu_{B_1}(v_0)$$

$$\alpha_2 = \mu_{A_2}(u_0) \wedge \mu_{B_2}(v_0)$$

where $\mu_{A_1}(u_0)$ and $\mu_{B_2}(v_0)$ are the degrees of partial match between the user-supplied data (u_0 and v_0) and the data (A_i and B_i) in the rule base.

(2)   Fuzzy input

If the input are given as fuzzy sets A' and B', the matching degrees α_i of rules are

$$\alpha_i = \min[\max_u(\mu_{A'}(u) \wedge \mu_{A_i}(u)), \max_v(\mu_{B'}(v) \wedge \mu_{B_i}(v))]$$

for i = 1, 2

We see that the matching degrees are obtained through the minimum operation in the Cartesian product. These relations play a key role in the following four inference methods.

9.4.1 Mamdani Method

This method uses the minimum operation R_C as a fuzzy implication and the max-min operator for the composition. Let's suppose a rule base is given in the following form.

$\quad\quad$ R_i: if u is A_i and v is B_i then w is C_i, $\quad$ i = 1, 2, ..., n

$\quad\quad\quad$ for $u \in$ U, $v \in$ V, and $w \in$ W.

then, $R_i = (A_i \text{ and } B_i) \rightarrow C_i$ is defined by

$$\mu_{R_i} = \mu_{(A_i \text{ and } B_i \rightarrow C_i)}(u, v, w)$$

(1) $\quad$ When input data are singleton $u = u_0$, $v = v_0$

$$\mu_{C_i'}(w) = [\mu_{A_i}(u_0) \text{ and } \mu_{B_i}(v_0)] \rightarrow \mu_{C_i}(w)$$

The Mamdani method uses the minimum operation ($\wedge$) for the fuzzy implication ($\rightarrow$). From lemma 5,

$$\mu_{C_i'}(w) = \alpha_i \wedge \mu_{C_i}(w)$$

$$where \; \alpha_i = \mu_{A_i}(u_0) \wedge \mu_{B_i}(v_0)$$

From the previous Lemma 3, we know the membership function μ_C of the inferred consequence C is given by the aggregated result derived from individual control rules. Thus, when there are two rules R_1 and R_2,

$$\mu_{C'}(w) = \mu_{C_1'} \vee \mu_{C_2'}$$

$$= [\alpha_1 \wedge \mu_{C_1}(w)] \vee [\alpha_2 \wedge \mu_{C_2}(w)]$$

The procedure of Mamdani fuzzy inference when the inputs are given as singletons is represented in (Fig 9.7).

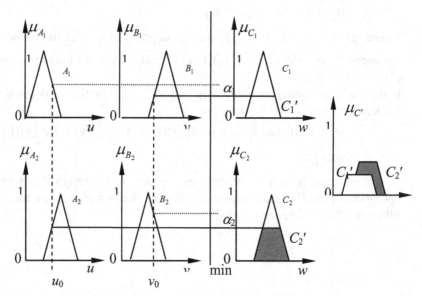

Fig. 9.7. Graphical representation of Mamdani method with singleton input

Therefore in general, from Lemma 3,

$$\mu_{C'}(w) = \bigvee_{i=1}^{n}[\alpha_i \wedge \mu_{C_i}(w)] = \bigvee_{i=1}^{n}\mu_{C'_i}(w)$$

$$C' = \bigcup_{i=1}^{n} C'_i$$

(2)   When input data are fuzzy sets, A' and B'
From Lemma 6,

$$\mu_{C'_i}(w) = \alpha_i \wedge \mu_{C_i}(w)$$

where $\alpha_i = \min[\max_u(\mu_{A'}(u) \wedge \mu_{A_i}(u)), \max_v(\mu_{B'}(v) \wedge \mu_{B_i}(v))]$

From Lemma 3, we have the aggregated result

$$\mu_{C'}(w) = \bigvee_{i=1}^{n}[\alpha_i \wedge \mu_{C_i}(w)] = \bigvee_{i=1}^{n}\mu_{C'_i}(w)$$

$$C' = \bigcup_{i=1}^{n} C'_i$$

The graphical interpretation of this inference is given in Fig 9.8.

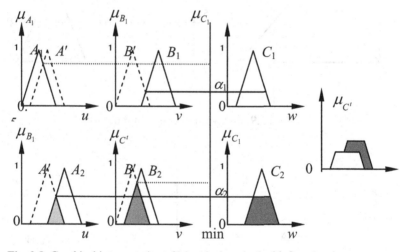

Fig. 9.8. Graphical interpretation of Mamdani method with fuzzy set input

The result C' is a fuzzy set and thus if we want to obtain a deterministic control action, a defuzzification method is used which will be discussed in the next chapter.

Example 9.5 There is a fuzzy rulebase including one rule such as :
                    R: If u is A then v is B
where A=(0, 2, 4) and B=(3, 4, 5) are triangular fuzzy sets.

If an input is given as singleton value $u_0=3$, how can we calculate the output B′ using the Mamdani method?

We can see the matching degree between A and u_0 is $\alpha=0.5$ Therefore the output B′ is obtained by the intersection between B and $\alpha=0.5$. That is, B′ is expressed by the lower area of 0.5 in B(Fig.9.9).

Now, consider the case that input is given as a triangular set A′=(0, 1, 2) . That is,

$$\mu_{A'}(x) = x \text{ for } 0 \leq x \leq 1$$
$$= -x+2 \text{ for } 1 \leq x \leq 2$$
$$= 0 \quad \text{otherwise}$$

We can obtain the mathching degree $\alpha=2/3$ and then B′ is the lower part of 2/3 in B (Fig 9.10).   □

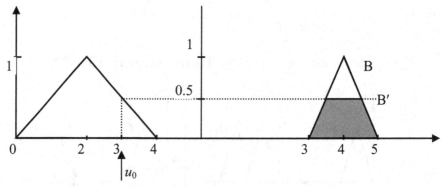

Fig. 9.9. Fuzzy inference with input $u_0=3$

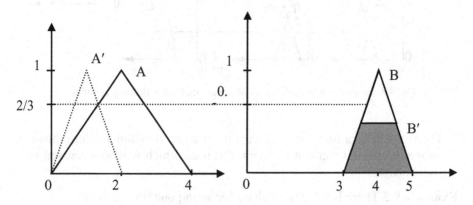

Fig. 9.10. Fuzzy inference with input A′=(0, 1, 2).

9.4.2 Larsen Method

This method uses the product operator R_P for the fuzzy implication and the max-product operator for the composition. For the following rule base,

R_i: if u is A_i and v is B_i then w is C_i, $\quad$ i = 1, 2, ... , n

then

$$R_i = (A_i \ and \ B_i) \rightarrow C_i \ \text{ is defined by}$$

$$\mu_{R_i} = \mu_{(A_i \ and \ B_i \rightarrow C_i)}(u,v,w)$$

(1) When the singleton input data are given as $u = u_0$, $v = v_0$, from Lemma 5 we have

$$\mu_{C_i'}(w) = [\mu_{A_i}(u_0) \ and \ \mu_{B_i}(v_0)] \rightarrow \mu_{C_i}(w)$$

$$= [\mu_{A_i}(u_0) \wedge \mu_{B_i}(v_0)] \cdot \mu_{C_i}(w)$$

$$= \alpha_i \cdot \mu_{C_i}(w) \quad where \ \alpha_i = \mu_{A_i}(u_0) \wedge \mu_{B_i}(v_0)$$

From Lemma 3,

$$\mu_{C'}(w) = \bigvee_{i=1}^{n}[\alpha_i \cdot \mu_{C_i}(w)] = \bigvee_{i=1}^{n} \mu_{C_i'}(w)$$

$$C' = \bigcup_{i=1}^{n} C_i'$$

The graphical representation of this method with singleton input is given in (Fig 9.9)

(2) When the input data are given as the form of fuzzy sets A' and B', from Lemma 6, we know

$$\mu_{C_i'}(w) = \alpha_i \cdot \mu_{C_i}(w)$$

$$where \ \alpha_i = \min[\max_u(\mu_{A'}(u) \wedge \mu_{A_i}(u)), \max_v(\mu_{B'}(v) \wedge \mu_{B_i}(v))$$

From Lemma 3, we have

$$\mu_{C'}(w) = \bigvee_{i=1}^{n}[\alpha_i \cdot \mu_{C_i}(w)] = \bigvee_{i=1}^{n} \mu_{C_i'}(w)$$

$$C' = \bigcup_{i=1}^{n} C_i'$$

The graphical interpretation of this inference is shown in (Fig 9.10.)

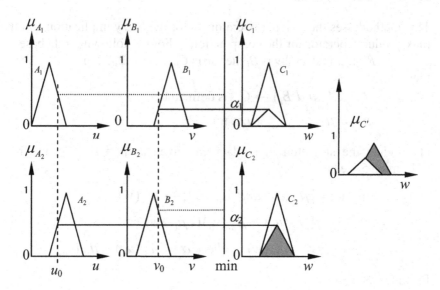

Fig. 9.11. Graphical representation of Larsen method with singleton input

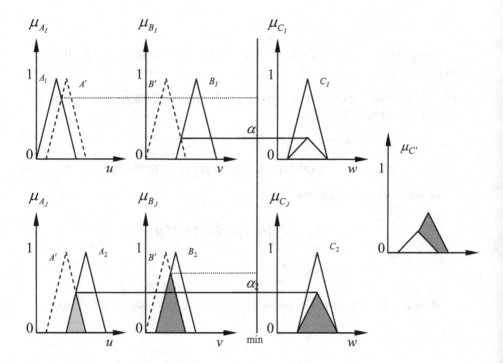

Fig. 9.12. Graphical representation of Larsen method with fuzzy set input

Example 9.6   There is a fuzzy rule
                    R : if u is A and v B then w is C
                    where   A=(0, 2, 4), B=(3, 4, 5) and C=(3, 4, 5)
i) Find inference result C' when input is u_0 =3, v_0=4 by using Larsen
   method
ii) Find inference result C' when input is A=(0, 1, 2) and B=(2, 3, 4).
   The solutions are illustrated in (Figs 9.13 , 9.14), respectively.

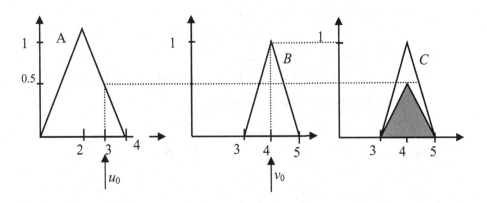

Fig. 9.13. Larsen method with input u_0=3, v_0=4

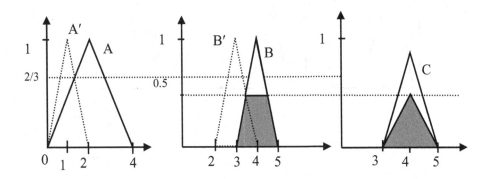

Fig. 9.14. Larsen method with input A'=(0, 1, 2), B'=(2, 3, 4).

9.4.3 Tsukamoto Method

In this method, the consequence of each fuzzy rule is represented by a fuzzy set with a monotonic membership function, as shown in (Fig 9.5.) The rule base has the form as:

R_i: if u is A_i and v is B_i, then w is C_i, $i = 1, 2, \dots, n$

where $\mu_{C_i}(w)$ is a monotonic function.

As a result, the inferred output of each rule is defined as a crisp value induced by the rule's matching degree (firing strength).   The overall output is taken as the weighted average of each rule's output.

We suppose that the set C_i has a monotonic membership function $\mu_{C_i}(w)$ and that α_i is the matching degree of ith rule.

(1)   For the singleton input (u_0, v_0)

$$\alpha_i = \mu_{A_i}(u_0) \wedge \mu_{B_i}(v_0)$$

(2)   For the fuzzy set input (A', B')

$$\alpha_i = \min[\max_u(\mu_{A'}(u) \wedge \mu_{A_i}(u)), \max_v(\mu_{B'}(v) \wedge \mu_{B_i}(v))]$$

Then the result of ith rule is obtained by

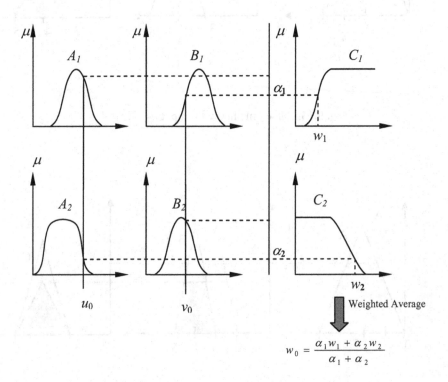

$$w_0 = \frac{\alpha_1 w_1 + \alpha_2 w_2}{\alpha_1 + \alpha_2}$$

Fig. 9.15. Graphical representation of Tsukamoto method

$$w_i = \mu_{C_i}^{-1}(\alpha_i)$$

The final result is derived from the weighted average like in the following when there are two rules.

$$w_0 = \frac{\alpha_1 w_1 + \alpha_2 w_2}{\alpha_1 + \alpha_2}$$

Since each rule infers a crisp result, the Tsukamoto fuzzy model aggregates each rule's output by the weighted average method. Therefore, it avoids the time-consuming process of defuzzification.

9.4.4 TSK Method

This method was proposed by Takagi, Sugeno, and Kang. A typical fuzzy rule in this model has the form

If u is A and v is B then $w = f(u, v)$,

A and B are fuzzy sets in the antecedent while $w = f(u, v)$ is a crisp function in the consequent. Usually $f(u, v)$ is a polynomial in the input variable u and v, and thus this method works when inputs are given as singleton values (Fig 9.16).

For simplicity, assume we have two fuzzy rules as follows.

R_1: if u is A_1 and v is B_1 then $w = f_1(u, v) = p_1u + q_1v + r_1$
R_2: if u is A_1 and v is B_1 then $w = f_2(u, v) = p_2u + q_2v + r_2$
where p_1, p_2, q_1, and q_2 are constant.

The inferred value of the control action from the first rule is $f_1(u_0, v_0)$ where u_0 and v_0 are singleton inputs, and α_1 is the matching degree. The inferred value from the second is $f_2(u, v)$ with the matching degree α_2. The matching degrees are obtained like in the previous methods.

$$\alpha_i = \mu_{A_i}(u_0) \wedge \mu_{B_i}(v_0)$$

They are all crisp values. The aggregated result is given by the weighted average.

$$w_0 = \frac{\alpha_1 f_1(u_0, v_0) + \alpha_2 f_2(u_0, v_0)}{\alpha_1 + \alpha_2}$$

$$= \frac{\alpha_1 w_1 + \alpha_2 w_2}{\alpha_1 + \alpha_2}$$

This method also saves the defuzzification time because the final result w_0 is a crisp value.

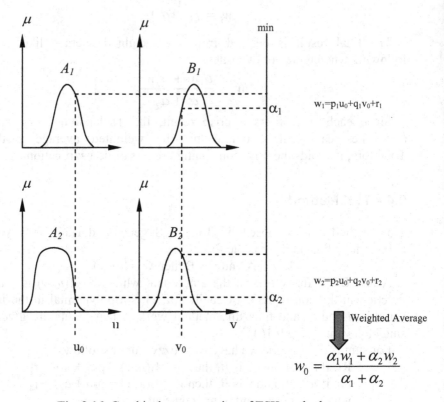

Fig. 9.16. Graphical representation of TSK method

[SUMMARY]

☐ Composition
  – composition of crisp sets (Cartesian product)
  – composition of fuzzy sets
  – composition of crisp relations
  – composition of fuzzy relations

☐ Composition of fuzzy sets (Fuzzy Cartesian product)
  – Fuzzy conjunction
  – Fuzzy disjunction

☐ Composition of fuzzy relations
  – max-min compositions
  – max-product composition

☐ Fuzzy rule
  – Form: if x is A, then y is B
  – Fuzzy implication: $A \rightarrow B$
  – Relation R(x, y): $A \times B$

☐ Interpretation of fuzzy implication
  – Min operation: R_C
  – Product operation: R_P

☐ Decomposition of rule base
  – Fuzzy rule base $R = \{ \overset{n}{\underset{i=1}{U}} R^i_{MIMO} \}$

  – $\{ \overset{n}{\underset{i=1}{U}} R^i_{MIMO} \} = \{RB^1_{MISO}, RB^2_{MISO}, \cdots, RB^q_{MISO}\}$

  – $R^i_{MIMO} : (A_i \times \cdots \times B_i) \rightarrow (z_1 + \cdots + z_q)$

  – $R^k_{MISO} : (A_i \times \cdots \times B_i) \rightarrow z_k$

☐ Compositional rule of inference
  – Implication R: $A \rightarrow B$, $R = A \times B$
  – Composition operator: $B' = A' \circ R$

☐ Mamdani implication and composition
  – Implication: min operator

- Composition: max-min

☐ Larsen implication and composition
  - Implication: algebraic product
  - Composition: max-product

☐ Inference with fuzzy rule base
  - Lemma 1: For 1 singleton input, result C' is obtained from C and matching degree α_1.

$$\mu_{C'}(w) = \alpha_1 \wedge \mu_C(w)$$
$$\text{where } \alpha_1 = \mu_A(u_0)$$

  - Lemma 2: For 1 fuzzy input, result C' is obtained from C and matching degree α_1.

$$\mu_{C'}(w) = \alpha_1 \wedge \mu_C(w)$$
$$\text{where } \alpha_1 = \max_u [\mu_{A'}(u) \wedge \mu_A(u)]$$

  - Lemma 3: Total result C' is an aggregation of individual result C_i'.

$$C' = (A', B') \circ \overset{n}{\underset{i=1}{U}} R_i = \overset{n}{\underset{i=1}{U}} (A', B') \circ R_i$$

  - Lemma 4: $(A_i \times B_i \rightarrow C_i)$ consists of $(A_i \rightarrow C_i)$ and $(B_i \rightarrow C_i)$.

$$C_i' = (A', B') \circ (A_i \text{ and } B_i \rightarrow C_i)$$
$$= [A' \circ (A_i \rightarrow B_i)] \cap [B' \circ (B_i \rightarrow C_i)]$$

  - Lemma 5: For singleton input, C_i' is from α_i and C_i.

$$\mu_{C_i'}(w) = \alpha_i \wedge \mu_{C_i}(w) \text{ for } R_C$$
$$\text{where } \alpha_i = \min[\mu_{A_i}(u_0), \mu_{B_i}(v_0)]$$

  - Lemma 6: For fuzzy input, C_i' is from α_i and C_i

$$\mu_{C_i'}(w) = \alpha_i \wedge \mu_{C_i}(w) \text{ for } R_C$$
$$\text{where } \alpha_i = \min[\max_u(\mu_{A'}(u) \wedge \mu_A(u)), \max_v(\mu_{B'}(v), \mu_B(v))]$$

☐ Inference method
  - Mamdani method: R_C
    • singleton input
    • fuzzy set input
  - Larsen method: R_P
    • singleton input
    • fuzzy set input
  - Tsukamoto method
    • The membership function of fuzzy set in the consequent is a monotonic function.
    • Final result is a weighted average of individual rules' outputs

- TSK method
  - The consequent part is a crisp function of input variables.
  - Final result is a weighted average of individual rules' outputs.

[EXERCISES]

9.1 Explain the following compositions with examples
   a) Composition of crisp sets
   b) Composition of fuzzy sets
   c) Composition of crisp relations
   d) Composition of fuzzy relations

9.2 Enumerate and explain the operators for the composition of fuzzy relations

9.3 There is a fuzzy rule such as
$$\text{If } x \text{ is A and } y \text{ is B then } z \text{ is C}$$
Define the relation $R(x, y, z)$ of the implication $A \times B \to C$ with fuzzy implication operators (min and product operators). You can define it by giving the membership function $\mu_R(x, y, z)$ of the relation.

9.4 There are a fuzzy rule and fuzzy sets.
   R: If x is A then y is C.
   or   R: $A(x) \to C(y)$

A	a_1	a_2	a_3	a_4
μ_A	0.2	0.4	0.6	0.8

C	y_1	y_2	y_3
μ_C	0.4	0.6	0.9

Calculate the implication relations $R(x, y)$ by using the min and product operators.

9.5 There is a fact A' given for the rule in the above exercise.

A'	a_1	a_2	a_3	a_4
$\mu_{A'}$	0.5	0.6	0.7	1.0

Calculate the output C' when you apply composition operations to the fact A' and the rule.$R(x,y)$

9.6 Prove the following equation and show its interpretation.
$$R = \{ \overset{n}{\underset{i=1}{U}} R^i_{MIMO} \} = \{ RB^1_{MISO}, RB^2_{MISO}, \cdots, RB^q_{MISO} \}$$

9.7 What is the compositional rule of inference?

9.8 Introduce operators which you can apply for the compositional rule of inference.

9.9 Explain meanings of the 5 Lemmas in this chapter.

9.10 What is the matching degree of inference?   Define the matching degree α in the following cases
   a) R_i: $A_i \rightarrow C_i$; singleton input u_0 is given.
   b) R_i: $A_i \rightarrow C_i$; fuzzy set input A' is given.
   c) R_i: $A_i \times B_i \rightarrow C_i$; singleton inputs u_0 and v_0 are given.
   d) R_i: $A_i \times B_i \rightarrow C_i$; fuzzy set inputs A' and B' are given.

9.11 There are the following rules and fuzzy set inputs A' and B'.

$$R : \overset{n}{\underset{i=1}{U}} R_i$$

$$R_i : A_i \text{ and } B_i \rightarrow C_i$$

Prove the following equation.

$$\mu_{C'}(w) = [\mu_{A'}(u), \mu_{B'}(v)] \circ \max_{u,v,w}[\mu_{R_1}(u, v, w), \mu_{R_2}(u, v, w), \cdots, \mu_{R_n}(u, v, w)]$$

$$= \max_{u,v,w}\{[(\mu_{A'}(u), \mu_{B'}(v)) \circ \mu_{R_1}(u, v, w)],$$

$$[(\mu_{A'}(u), \mu_{B'}(v)) \circ \mu_{R_2}(u, v, w)],$$

$$\cdots$$

$$[(\mu_{A'}(u), \mu_{B'}(v)) \circ \mu_{R_n}(u, v, w)]\}$$

9.12 Show contributions of each Lemma in the chapter to the Mamdani inference method.

9.13 What is the difference between Tsukamoto method and TSK method.

9.14 Can we use the same matching degree α_i for Mamdani, Larsen, and Tsukamoto methods when there are same rule base and inputs?

9.15 There is a fuzzy rulebase with only one rule:
      R: if x is A and y is b then z is C,
      where A=(0,1,2), B=(1,2,3) and C=(5,6,7).
      are triangular fuzzy sets.
   a) Calculate the output fuzzy set when input is given as x_0=1 and y_0=1.5.

b) Find the output fuzzy set when input is given as
$$A'=(1,2,3) \quad \text{and} \quad B'=(1.5,2.5,3.5)$$

Chapter 10. FUZZY CONTROL AND FUZZY EXPERT SYSTEMS

The fuzzy logic controller (FLC) is introduced in this chapter. After introducing the architecture of the FLC, we study its components step by step and suggest a design procedure of the FLC. An example of the design procedure is also given. The structure and function of the fuzzy expert systems are similar to those of the FLC, and thus the design procedure of FLC can be used in the fuzzy expert systems.

10.1 Fuzzy Logic Controller

10.1.1 Advantage of Fuzzy Logic Controller

Fuzzy logic is much closer in spirit to human thinking and natural language than the traditional (classical) logical systems. Basically, it provides an effective means of capturing the approximate, inexact nature of the real world. Therefore, the essential part of the fuzzy logic controller (FLC) is a set of linguistic control strategy based on expert knowledge into an automatic control strategy.

The FLC is considered as a good methodology because it yields results superior to those obtained by conventional control algorithms. In particular the FLC is useful in two cases.

(1)   The control processes are too complex to analyze by conventional quantitative techniques.

(2)   The available sources of information are interpreted qualitatively, inexactly, or uncertainly.

Indeed, the advantage of FLC can be summarized as follows.

(1)   Parallel or distributed control: in the conventional control system, a control action is determined by single control strategy like $\mu = f(x_1, x_2, \dots, x_n)$. But in FLC, the control strategy is represented by multiple fuzzy rules, and thus it is easy to represent complex systems and nonlinear systems.

(2) Linguistic control: the control strategy is modeled by linguistic terms and thus it is easy to represent the human knowledge.

(3) Robust control: there are more than one control rule and thus, in general, one error of a rule is not fatal for the whole system.

10.1.2 Configuration of Fuzzy Logic Controller

There is no systematic procedure for the design of an FLC. However we can present here a basic configuration of FLC as shown in (Fig 10.1.). The configuration consists of four main components: fuzzification interface, knowledge base, decision-making logic, and defuzzification interface.

(1) The fuzzification interface transforms input crisp values into fuzzy values and it involves the following functions.
- Receives the input values
- Transforms the range of values of input variable into corresponding universe of discourse
- Converts input data into suitable linguistic values (fuzzy sets).

This component is necessary when input data are fuzzy sets in the fuzzy inference.

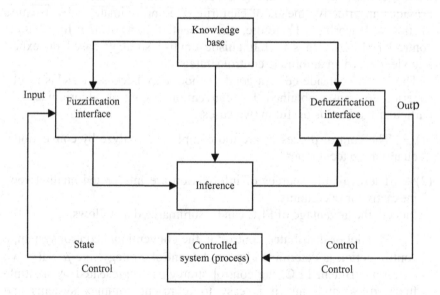

Fig. 10.1. Configuration of FLC

(2)   The knowledge base contains a knowledge of the application domain and the control goals.   It consists of a data base and a linguistic rule base.
- The data base contains necessary definitions which are used in control rules and data manipulation.
- The linguistic rule base defines the control strategy and goals by means of linguistic control rules.

(3)   The decision-making logic performs the following functions
- simulates the human decision-making procedure based on fuzzy concepts
- Infers fuzzy control actions employing fuzzy implication and linguistic rules.

(4)   The defuzzification interface the functions
- A scale mapping which converts the range of output values into corresponding universe of discourse
- Defuzzification which yields a nonfuzzy control action from an inferred fuzzy control action.

In the following sections, the control components will be developed in detail.

10.1.3 Choice of State Variables and Control Variables

Before starting the detailed procedure of the FLC design, we have to choose the variables.   A fuzzy control system is designed to control a pr ocess, and thus it is needed to determine state variables and control variables of the process.   The state variables become input variables of the fuzzy control system, and the control variables become output variables.   Selection of the variables depends on expert knowledge on the process.   In particular, variables such as state, state error, state error deviation, and state error integral are often used.

10.2 Fuzzification Interface Component

In the fuzzification component, there are three main issues to be considered: scale mapping of input data, strategy for noise and selection of fuzzification functions.

(1)   Scale mapping of input data: We have to decide a strategy to convert the range of values of input variables into corresponding universe of discourse. When an input value is come through a measuring system, the values must be located in the range of input variables.   For example, if the range of input variables was normalized between -1

and +1, a procedure is needed which maps the observed input value into the normalized range.

(2)   Strategy for noise: When observed data are measured, we may often think that the data were disturbed by random noise. In this case, a fuzzification operator should convert the probabilistic data into fuzzy numbers. In this way, computational efficiency is enhanced since fuzzy numbers are much easier to manipulate than random variables. Otherwise, we assume that the observed data do not contain vagueness, and then we consider the observed data as a fuzzy singleton. A fuzzy singleton is a precise value and hence no fuzziness is introduced by fuzzification in this case. In control applications, the observed data are usually crisp and used as fuzzy singleton inputs in the fuzzy reasoning.

(3)   Selection of fuzzification function: A fuzzification operator has the effect of transforming crisp data into fuzzy sets.

$$x = \text{fuzzifier}\,(x_0)$$

Where x_0 is a observed crisp value and x is a fuzzy set, and fuzzifier represents a fuzzification operator.

(Fig 10.1) shows a fuzzification function which transforms crisp data into a fuzzy singleton value.

(Fig 10.2) shows a fuzzification function transforming a crisp value into a triangular fuzzy number. The peak point of this triangle corresponds to the mean value of a data set, while the base is twice the standard deviation of the data set.

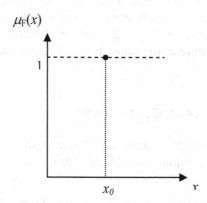

Fig. 10.2. Fuzzification function for fuzzy singleton

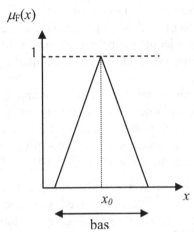

Fig. 10.3. Fuzzification function for fuzzy triangular number

10.3 Knowledge Base Component

10.3.1 Data Base

The knowledge base of an FLC is comprised of two parts: a data base and a fuzzy control rule base.   We will discuss some issues relating to the data base in this section and the rule base in the next section.

In the data base part, there are four principal design parameters for an FLC: discretization and normalization of universe of discourse, fuzzy partition of input and output spaces, and membership function of primary fuzzy set.

(1) Discretization and normalization of universe of discourse
The modeling of uncertain information with fuzzy sets raises the problem of quantifying such information for digital computers.   A universe of discourse in an FLC is either discrete or continuous.   If the universe is continuous, a discrete universe may be formed by a discretization procedure.   A data set may be also normalized into a certain range of data.

1).   Discretization of a universe of discourse: It is often referred to as quantization.   The quantization discretizes a universe into a certain number of segments.   Each segment is labeled as a generic element and forms a discrete universe. A fuzzy set is then defined on the discrete universe of discourse. The number of quantization levels affects an important influence on the control performance, and thus it should be large enough to give adequate approximation. That number should be determined in considering both the control quality and the memory storage in computer.

For the discretization, we need a scale mapping, which serves to transform measured variables into values in the discretized universe. The mapping can be uniform (linear), nonuniform (nonlinear), or both. (Table 10.1) shows an example of discretization, where a universe of discourse is discretized into 13 levels (-6, -5, -4, ... , 0, 1, ... , 5, 6).

2). Normalization of a universe of discourse: It is a discretization into a normalized universe. The normalized universe consists of finite number of segments. The scale mapping can be uniform, nonuniform, or both. (Table 10.2) shows an example, where the universe of discourse [-6.9, +4.5] is transformed into the normalized closed interval [-1, 1].

Table 10.1 An example of discretization

Range	Level No.
$x \leq -2.4$	-6
$-2.4 < x \leq -2.0$	-5
$-1.6 < x \leq -0.8$	-4
$-0.8 < x \leq -0.4$	-3
$-0.4 < x \leq -0.2$	-2
$-0.2 < x \leq -0.1$	-1
$-0.1 < x \leq +0.1$	0
$+0.1 < x \leq +0.2$	1
$+0.2 < x \leq +0.4$	2
$+0.4 < x \leq +0.8$	3
$+0.8 < x \leq +1.1$	4
$+1.1 < x \leq +1.4$	5
$+1.4 < x$	6

Table 10.2. An example of normalization

Range	Normalized segments	Normalized universe
[−6.9, −4.1]	[−1.0, −0.5]	
[−4.1, −2.2]	[−0.5, −0.3]	
[−2.2, −0.0]	[−0.3, 0.0]	[−1.0, +1.0]
[−0.0, +1.0]	[0.0, +0.2]	
[+1.0, +2.5]	[+0.2, +0.6]	
[+2.5, +4.5]	[+0.6, +1.0]	

3). Normalization of a universe of discourse: It is a discretization into a normalized universe. The normalized universe consists of finite number of segments. The scale mapping can be uniform, nonuniform, or both. (Table 10.2) shows an example, where the universe of discourse [-6.9, +4.5] is transformed into the normalized closed interval [-1, 1].

(2) Fuzzy partition of input and output spaces
A linguistic variable in the antecedent of a rule forms a fuzzy input space, while that in the consequent of the rule forms a fuzzy output space. In general, a linguistic variable is associated with a term set. A fuzzy partition of the space determines how many terms should exist in a term set. This is the same problem to find the number of primary fuzzy sets (linguistic terms).
There are seven linguistic terms often used in the fuzzy inference:

NB: negative big

NM: negative medium

NS: negative small

ZE: zero

PS: positive small

PM: positive medium

PB: positive big

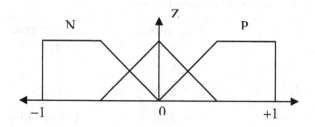

(a) N: negative, Z: zero, P: positive

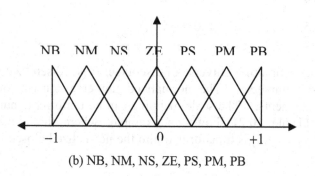

(b) NB, NM, NS, ZE, PS, PM, PB

Fig. 10.3. Example of fuzzy partition with linguistic terms

A typical example is given in (Fig 10.3) representing two fuzzy partition in the same normalized universe [−1, +1]. Membership functions with triangle and trapezoid shapes are used here.

The number of fuzzy terms in a input space determines the maximum number of fuzzy control rules. Suppose a fuzzy control system with two input and one output variables. If the input variables have 5 and 4 terms, the maximum number of control rules that we can construct is 20 (5 × 4) as shown in (Fig 10.4.). (Fig 10.5) shows an example of system having 3 fuzzy rules.

A fuzzy control system could always infer a proper control action for every state of process. This property is concerned with the supports on which primary fuzzy sets are defined. The union of these supports should cover the related universe of discourse in relation to some level. For example, in (Fig 10.3), any input value is included to at least one linguistic term with membership value greater than 0.5.

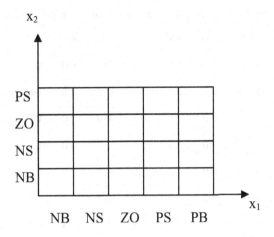

Fig. 10.4. A fuzzy partition in 2-dimension input space

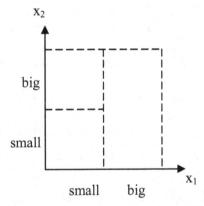

Fig. 10.5. A fuzzy partition having three rules

(3)   Membership function of primary fuzzy set
There are various types of membership functions such as triangular, trapezoid, and bell shapes.   (Table 10.3) shows an example defining triangular membership functions on the discretized universe of discourse in (Table 10.1.)   For example, term NM is defined such as:

$$\mu_{NM}(-6) = 0.3$$
$$\mu_{NM}(-5) = 0.7$$
$$\mu_{NM}(-4) = 1.0$$
$$\mu_{NM}(-3) = 0.7$$
$$\mu_{NM}(-2) = 0.3$$

Table 10.3. Definition of triangular membership function

Level No.	NB	NM	NS	ZE	PS	PM	PB
−6	1.0	0.3	0.0	0.0	0.0	0.0	0.0
−5	0.7	0.7	0.0	0.0	0.0	0.0	0.0
−4	0.3	1.0	0.3	0.0	0.0	0.0	0.0
−3	0.0	0.7	0.7	0.0	0.0	0.0	0.0
−2	0.0	0.3	1.0	0.3	0.0	0.0	0.0
−1	0.0	0.0	0.7	0.7	0.0	0.0	0.0
0	0.0	0.0	0.3	1.0	0.3	0.0	0.0
1	0.0	0.0	0.0	0.7	0.7	0.0	0.0
2	0.0	0.0	0.0	0.3	1.0	0.3	0.0
3	0.0	0.0	0.0	0.0	0.7	0.7	0.0
4	0.0	0.0	0.0	0.0	0.3	1.0	0.3
5	0.0	0.0	0.0	0.0	0.0	0.7	0.7
6	0.0	0.0	0.0	0.0	0.0	0.3	1.0

An example of bell shaped membership function is given in (Table 10.4) and (Fig 10.6), where fuzzy sets are defined on the normalized universe of discourse [−1, +1] given in (Table 10.2.)   They have the shapes of function of parameter mean m_f and standard deviation σ_f.

$$\mu_f(x) = \exp\{\frac{-(x-m_f)^2}{2\sigma_f^2}\}$$

10.3.2 Rule Base

A fuzzy system is characterized by a set of linguistic statements usually represented by in the form of "if-then" rules.   In this section, we examine several topics related to fuzzy control rules.

Table 10.4. Definition of bell-shaped membership function

Normalized universe	Normalized segments	m_f	σ_f	Fuzzy sets
	[−1.0, −0.5]	−1.0	0.4	NB
	[−0.5, −0.3]	−0.5	0.2	NM
	[−0.3, −0.0]	−0.2	0.2	NM
[−1.0, +1.0]	[−0.0, +0.2]	0.0	0.2	ZE
	[+0.2, +0.6]	0.2	0.2	PS
	[+0.6, +0.8]	0.5	0.2	PM
	[+0.8, +1.0]	1.0	0.4	PB

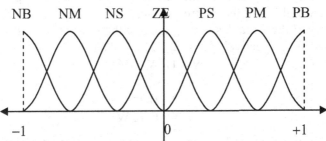

Fig. 10.6. Example of bell-shaped membership function

(1) Source of fuzzy control rules

There are two principal approaches to the derivation of fuzzy control rules. The first is a heuristic method in which rules are formed by analyzing the behavior of a controlled process. The derivation relies on the qualitative knowledge of process behavior. The second approach is basically a deterministic method which can systematically determine the linguistic structure of rules.

We can use four modes of derivation of fuzzy control rules. These four modes are not mutually exclusive, and it is necessary to combine them to obtain an effective system.

- Expert experience and control engineering knowledge: operating manual and questionnaire.
- Based on operators' control actions: observation of human controller's actions in terms of input-output operating data.
- Based on the fuzzy model of a process: linguistic description of the dynamic characteristics of a process.

– Based on learning: ability to modify control rules such as self-organizing controller.

(2)   Types of fuzzy control rules

There are two types of control rules: state evaluation control rules and object evaluation fuzzy control rules.

1). State evaluation fuzzy control rules: State variables are in the antecedent part of rules and control variables are in the consequent part.   In the case of MISO (multiple input single output), they are characterized as a collection of rules of the form.

$$R_1: \text{if } x \text{ is } A_1, \dots \text{ and } y \text{ is } B_1 \text{ then } z \text{ is } C_1$$
$$R_2: \text{if } x \text{ is } A_2, \dots \text{ and } y \text{ is } B_2 \text{ then } z \text{ is } C_2$$
$$\dots$$
$$R_n: \text{if } x \text{ is } A_n, \dots \text{ and } y \text{ is } B_n \text{ then } z \text{ is } C_n$$

where $x, \dots y$ and z are linguistic variables representing the process state variable and the control variable. $A_i, \dots B_i$ and C_i are linguistic values of the variables $x, \dots y$ and z in the universe of discourse $U, \dots$ V and W, respectively $i = 1, 2, \dots, n$.   That is,

$$x \in U, A_i \subset U$$
$$\dots$$
$$y \in V, B_i \subset V$$
$$z \in W, C_i \subset W$$

In a more general version, the consequent part is represented as a function of the state variable $x, \dots y$.

$$R_i: \text{if } x \text{ is } A_i, \dots \text{ and } y \text{ is } B_i \text{ then } z = f_i(x, \dots y)$$

The state evaluation rules evaluate the process state (e.g. state, state error, change of error) at time t and compute a fuzzy control action at time t.

In the previous section concerned with the fuzzy partition of input space, we said that the maximum number of control rules is defined by the partition.   In the input variable space, the combination of input linguistic term may give a fuzzy rule.   When there is a set of fuzzy rules as follows

$$R_i: \text{if } x \text{ is } A_i, \text{ and } y \text{ is } B_i \text{ then } z \text{ is } C_i$$
$$i = 1, 2, \dots, n$$

the rules can be represented as the form of table in (Fig 10.7.)

2). Object evaluation fuzzy control rules: It is also called predictive fuzzy control.   They predict present and future control acti -ons, and evaluate control objectives.   A typical rule is descri-bed as

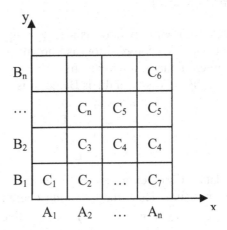

Fig 10.7. Fuzzy rules represented by a rule table

R_1: if (z is C_1 → (x is A_1 and y is B_1)) then z is C_1.
R_2: if (z is C_2 → (x is A_2 and y is B_2)) then z is C_2.

...

R_n: if (z is C_n → (x is A_n and y is B_n)) then z is C_n.

A control action is determined by an objective evaluation that satisfies the desired states and objectives. x and y are performance indices for the evaluation and z is control command. A_i and B_i are fuzzy values such as NM and PS. The most likely control rule is selected through predicting the results (x, y) corresponding to every control command C_i, $i = 1, 2, \ldots, n$.

In linguistic terms, the rule is interpreted as: if the performance index x is A_i and index y is B_i when a control command z_i is C_i, then this rule is selected, and the control command C_i is taken to be the output of the controller.

10.4 Inference (Decision Making Logic)

In general, in decision making logic part, we use four inference methods described in the previous chapter: Mamdani method, Larsen method, Tsukamoto method, and TSK method.

10.4.1 Mamdani Method

This method uses minimum operator as a fuzzy implication operator, and max-min operator for the composition as shown in (sec 9.4.)   Suppose fuzzy rules are given in the following form.

$$R_i: \text{if } x \text{ is } A_i \text{ and } y \text{ is } B_i \text{ then } z \text{ is } C_i,$$
$$i = 1, 2, \dots, n$$
$$x \in U, A_i \subset U$$
$$y \in V, B_i \subset V$$
$$z \in W, C_i \subset W$$

(1) When input data are singleton such as $x = x_0$ and $y = y_0$ (in this case, the input data are not fuzzified.), the matching degrees (firing strength) of A_i and B_i are $\mu_{A_i}(x_0)$ and $\mu_{B_i}(y_0)$, respectively.   Therefore the matching degree of rule R_i is

$$\alpha_i = \mu_{A_i}(x_0) \wedge \mu_{B_i}(y_0)$$

then $\mu_{C_i'}(z) = \alpha_i \wedge \mu_{C_i}(z)$ where C_i' is the result of rule R_i.   The aggregated result C' derived from individual control rules is defined as follows:

$$\mu_{C'}(z) = \overset{n}{\underset{i=1}{V}}[\alpha_i \wedge \mu_{C_i}(z)]$$

$$C' = \overset{n}{\underset{i=1}{U}} C_i'$$

(2) When input data are fuzzy sets, A' and B'
$$\alpha_i = \min[\underset{x}{\max}(\mu_{A'}(x) \wedge \mu_A(x)), \underset{v}{\max}(\mu_{B'}(y) \wedge \mu_B(y))]$$
$$\mu_{C_i'}(z) = \alpha_i \wedge \mu_{C_i}(z)$$
$$i = 1, 2, \cdots, n$$
The aggregate result C' is defined by
$$\mu_{C'}(z) = \overset{n}{\underset{i=1}{V}}[\alpha_i \wedge \mu_{C_i}(z)]$$
$$C' = \overset{n}{\underset{i=1}{U}} C_i'$$

10.4.2 Larsen Method

This method uses the product operator (.) for the fuzzy implication, and the max-product operator for the composition.   Suppose fuzzy rules are given in the following form.

$$R_i: \text{if } x \text{ is } A_i \text{ and } y \text{ is } B_i \text{ then } z \text{ is } C_i,$$

$$i = 1, 2, \ldots, n$$

(1) When input data are singleton, $x = x_0$ and $y = y_0$. The matching degrees is

$$\alpha_i = \mu_{A_i}(x_0) \wedge \mu_{B_i}(y_0)$$

the result C_i' of rule R_i is defined by

$$\mu_{C_i'}(z) = \alpha_i \cdot \mu_{C_i}(z)$$

the aggregated result C' is

$$\mu_{C'}(z) = \bigvee_{i=1}^{n} [\alpha_i \cdot \mu_{C_i}(z)]$$

or

$$C' = \overset{n}{\underset{i=1}{U}} C_i'$$

(2) When input data are given as the form of fuzzy sets, A' and B', we have matching degrees as

$$\alpha_i = \min[\max_{x}(\mu_{A'}(x) \wedge \mu_{A}(x)), \max_{y}(\mu_{B'}(y) \wedge \mu_{B}(y))]$$

The result C_i' of rule R_i is defined by

$$\mu_{C_i'}(z) = \alpha_i \cdot \mu_{C_i}(z)$$

The aggregate result C' is

$$\mu_{C'}(z) = \bigvee_{i=1}^{n} [\alpha_i \wedge \mu_{C_i}(z)]$$

or

$$C' = \overset{n}{\underset{i=1}{U}} C_i'$$

10.4.3 Tsukamoto Method

This method is used when the consequent part of each rule is represented by fuzzy set with a monotonic membership function. The inferred output of each rule is defined as a crisp value induced by the rule's matching degree (firing strength).

We suppose fuzzy rules are given in the following form and the set C_i has a monotonic membership function $\mu_{C_i}(z)$

$$R_i: \text{if } x \text{ is } A_i \text{ and } y \text{ is } B_i \text{ then } z \text{ is } C_i,$$
$$i = 1, 2, \ldots, n$$

the matching degree α_i of each rule is defined like in the previous methods in the cases of both singleton input and fuzzy set input.

The result of z_i rule R_i is obtained by (Fig 10.8)

$$z_i = \mu_{C_i}^{-1}(\alpha_i)$$

The aggregated result z' is taken as the weighted average of each rule's output

$$z' = \frac{\alpha_1 z_1 + \alpha_2 z_2}{\alpha_1 + \alpha_2}$$

This method gives a crisp value as an aggregated result and thus there is no need to defuzzify it.

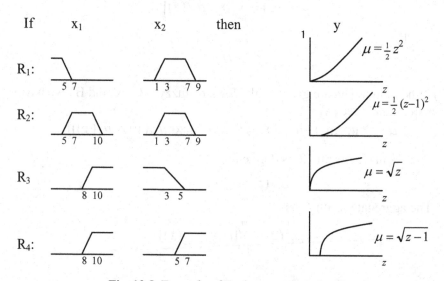

Fig. 10.8. Example of Tsukamoto control rules

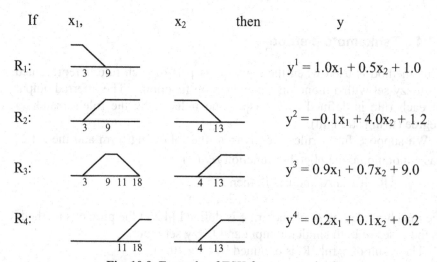

Fig. 10.9. Example of TSK fuzzy control rules

10.4.4 TSK Method

This method is used when the consequent part is given as a function of input variables.

$$R_i: \text{if } x \text{ is } A_i \text{ and } y \text{ is } B_i \text{ then } z \text{ is } f_i(x, y)$$

Where $z = f(x, y)$ is a crisp function of input variables x and y. Usually $f(x, y)$ has a polynomial form (Fig 10.9).

Suppose input data are singleton x_0 and y_0, then the inferred result of rule R_i is $f_i(x_0, y_0)$. The matching degree α_i of R_i is same with the previous one. Therefore the final aggregated result z' is the weighted average using the matching degree α_i

$$z' = \frac{\alpha_1 f_1(x_0, y_0) + \alpha_2 f_2(x_0, y_0)}{\alpha_1 + \alpha_2}$$

The final result is a crisp value and thus there is no need to defuzzify it.

10.5 Defuzzification

In many practical applications, a control command is given as a crisp value. Therefore it is needed to defuzzify the result of the fuzzy inference. A defuzzification is a process to get a non-fuzzy control action that best represents the possibility distribution of an inferred fuzzy control action. Unfortunately, we have no systematic procedure for choosing a good defuzzification strategy, and thus we have to select one in considering the properties of application case. The three commonly used strategies are described in this section.

10.5.1 Mean of Maximum Method (MOM)

The MOM strategy generates a control action which represents the mean value of all control actions, whose membership functions reach the maximum (Fig 10.10). In the case of a discrete universe, the control action may be expressed as

$$z_0 = \sum_{j=1}^{k} \frac{z_j}{k}$$

z_j: control action whose membership functions reach the maximum.

k: number of such control actions.

10.5.2 Center of Area Method (COA)

The widely used COA strategy generates the center of gravity of the possibility distribution of a fuzzy set C (Fig 10.11). In the case of a discrete universe, thus method gives

$$z_0 = \frac{\sum_{j=1}^{n} \mu_C(z_j) \cdot z_j}{\sum_{j=1}^{n} \mu_C(z_j)}$$

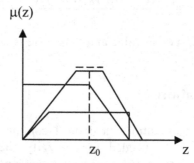

Fig. 10.10. Mean of maximum (MOM)

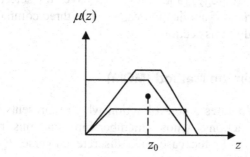

Fig. 10.11. Center of area (COA)

Where n is the number of quantization levels of the output, C is a fuzzy set defined on the output dimension (z).

10.5.3 Bisector of Area (BOA)

The BOA generates the action (z_0) which partitions the area into two regions with the same area (Fig 10.12).

$$\int_{\alpha}^{z_0} \mu_C(z)\, dz = \int_{z_0}^{\beta} \mu_C(z)\, dz$$

where $\alpha = \min\{z \mid z \in W\}$ $\beta = \max\{z \mid z \in W\}$

10.5.4 Lookup Table

Even with the many advantages, it is pointed out that the FLC has the problem of time complexity. It takes much time to compute the fuzzy inference and defuzzification. Therefore a lookup table is often used which simply shows relationships between input variables and control output actions. But the lookup table can be constructed after making the FLC and identifying the relationships between the input and output variables. In general, it is extremely difficult to get an acceptable lookup table of a nonlinear control system without constructing a corresponding FLC.E

Example 10.1 (Table 10.5.) shows an example of lookup table for the two input variables error (e) and change of error (ce), and control variable (v). The variables are all discretized and normalized in the range $[-1, +1]$. For example, when e = −1.0 and ce = −0.5, we can obtain v = −0.5 by using the lookup table instead of by executing the full fuzzy controller. □

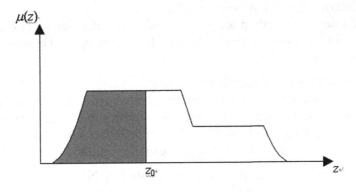

Fig. 10.12. Bisector of area (BOA)

Table 10.5. Example of lookup table

ce e	−1.0	−0.5	0	0.5	1.0
−1.0	−1.0	−0.5	−0.5	−0.5	−0.5
−0.5	−1.0	−1.0	0.5	−0.5	−0.5
0	−1.0	−1.0	0.5	−0.5	−0.5
0.5	0.5	0.5	1.0	1.0	1.0
1.0	0.5	0.5	0.5	1.0	1.0

10.6 Design Procedure of Fuzzy Logic Controller

When we decided to design a fuzzy logic controller, we can follow the following design procedure

(1) Determination of state variables and control variables
In general, the control variable is determined depending on the property of process to be controlled. But we have to select the state variables. In general, state, state error and error difference are often used. The state variables are input variables, and the control variables are output of our controller to be developed.

(2) Determination of inference method
We select one method among four inference methods described in the previous section. The decision is dependent upon the properties of process to be studied.

(3) Determination of fuzzification method
It is necessary to study the property of measured data of state variables. If there is uncertainty in the data, the fuzzification is necessary, and we have to select a fuzzification method and membership functions of fuzzy sets.   If there is no uncertainty, we can use singleton state variables.

(4) Discretization and normalization of state variable space
In general, it is useful to use discretized and normalized universe of discourse. We have to decide whether it is necessary and how we can do.

(5) Partition of variable space.
The state variables are input variables of our controller and thus the partition is important for the structure of fuzzy rules.   At this step,

partition of control space (output space of the controller) is also necessary.

(6) Determination of the shapes of fuzzy sets
It is necessary to determine the shapes of fuzzy sets and their membership functions for the partitioned input spaces and output spaces.

(7) Construction of fuzzy rule base
Now, we can build control rules. We determined the variables and corresponding linguistic terms in antecedent part and consequent part of each rule. The architecture of rules is dependent upon the inference method determined in step 2).

(8) Determination of defuzzification strategy
In general, we use singleton control values and thus we have to determine the method.

(9) Test and tuning
It is almost impossible to obtain a satisfactory fuzzy controller without tuning. In general it is necessary to verify the controller and tune it until when we get satisfactory results.

(10) Construction of lookup table
If the controller shows satisfactory performance, we have to decide whether we use a lookup table instead of using the inference system. The lookup table is often used to save computing the time of the inference and defuzzification. The lookup table shows the relationships between a combination of input variables and control actions.

10.7 Application Example of FLC Design

Servomotors are used in many automatic system including drivers for printers, floppy disks, tape recorders, and robot manipulations. The control of such servomotors is an important issue. The servomotor process shows nonlinear properties, and thus we apply the fuzzy logic control to the motor control. The task of the control is to rotate the shaft of the motor to a set point without overshout. The set point and process output in measured in degree.

(1) Determination of state variables and control variable

  1) State variables (input variable of controller):
    − Error equals the set point minus the process output (e).

- Change of error (ce) equals the error from the process output minus the error from the last process output.

2) Control variable (output variable of the controller):
   - Control input (v) equals the voltage applied to the process.

(2) Determination of inference method
   The Mamdani inference method is selected because it is simple to explain.

(3) Determination of fuzzification method
   We can measure the state variables without uncertainty and thus we use the measured singleton for the fuzzy inference

(4) Discretization and normalization
   The shaft encoder of the motor has a resolution of 1000.   The universes of discourse are as follows:

$$-1000 \le e \le 1000$$
$$-100 \le ce \le 100$$

The servo amplifier has an output range of 30 V and thus the control variables (v) are in the range

$$-30 \le v \le 30$$

We discretize and normalize the input variables in the range $[-1, +1]$ as shown in Table 10.6.   The control variable v is normalized in the range $[-1, +1]$ with the equation.

$$v' = \frac{1}{30} v$$

Table 10.6. Discretization and normalization

error (e)	error change (ce)	quantized level
$-1000 \le e \le -800$	$-100 \le ce \le -80$	-1.0
$-800 < e \le -600$	$-80 < ce \le -60$	-0.8
$-600 < e \le -400$	$-60 < ce \le -40$	-0.6
$-400 < e \le -200$	$-40 < ce \le -20$	-0.4
$-200 < e \le -100$	$-20 < ce \le -10$	-0.2
$-100 < e \le 100$	$-10 < ce \le 10$	0
$100 < e \le 200$	$10 < ce \le 20$	0.2
$200 < e \le 400$	$20 < ce \le 40$	0.4
$400 < e \le 600$	$40 < ce \le 60$	0.6
$600 < e \le 800$	$60 < ce \le 80$	0.8
$800 < e \le 1000$	$80 < ce \le 100$	1.0

(5) Partition of input space and output space

We partition space of each input and output variable into seven regions, and each region is associated with linguistic term as shown in (Fig 10.13.). Now we know the maximum number of possible fuzzy rules is 49.

(6) Determination of the shapes of fuzzy sets

We normalized the input and output variables on the same interval [−1, +1] and partitioned the region into seven subregions, and thus we define

(7) Partition of input space and output space

We partition space of each input and output variable into seven regions, and each region is associated with linguistic term as shown in (Fig 10.13.). Now we know the maximum number of possible fuzzy rules is 49.

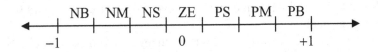

Fig. 10.13. Partition of space

(8) Determination of the shapes of fuzzy sets

We normalized the input and output variables on the same interval [−1, +1] and partitioned the region into seven subregions, and thus we define the primary triangular fuzzy sets for the all variables as shown in (Table 10.7.) and (Fig 10.14.)

Table 10.7. Definition of primary fuzzy sets

Level	NB	NM	NS	ZE	PS	PM	PB
−1	1	0.5	0	0	0	0	0
−0.8	0	1	0	0	0	0	0
−0.6	0	0.5	0.5	0	0	0	0
−0.4	0	0	1	0	0	0	0
−0.2	0	0	0.5	0.5	0	0	0
0	0	0	0	1	0	0	0
0.2	0	0	0	0.5	0.5	0	0
0.4	0	0	0	0	1	0	0
0.6	0	0	0	0	0.5	0.5	0
0.8	0	0	0	0	0	1	0
1.0	0	0	0	0	0	0.5	1

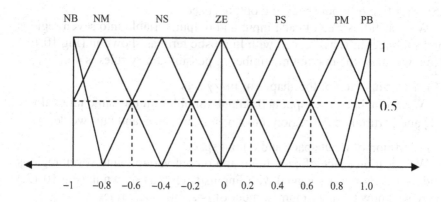

Fig. 10.14. Graphical representation of primary fuzzy sets

(9) Construction of fuzzy rules

We interviewed with an expert of the servomotor control, and we collect knowledge such as:

If the error is zero and the error change is positive small, then the control input is negative small.

This type of rules are rewritten in the following form

1) If e is PB and ce is any, then v is PB.

2) If e is PM and ce is NB, NM, or NS, then v is PS.

3) If e is ZE and ce is ZE, PS, or PM, then v is ZE.

4) If e is PS and ce is NS, ZE, or PS, then v is ZE.

5) If e is NS and ce is NS, ZE, PS, or PM, then v is NS

6) If e is NS or ZE and ce is PB, then v is PS.

The full set of fuzzy rules is summarized in the rule table in (Fig 10.15.)

(10) Determination of defuzzification strategy

We take the COA (center of area) method because it is most commonly used.

(11) Test and tuning

(12) Determination of defuzzification strategy

We take the COA (center of area) method because it is most commonly used.

(13) Test and tuning

We checked the performance of the developed controller and refined some fuzzy rules.

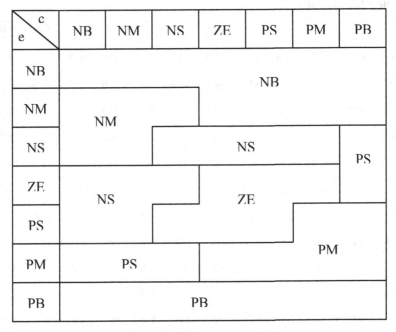

Fig. 10.15. Fuzzy rule table

(14) Construction of lookup table

After verifying the controller showing good performance, we decided to use a lookup table. We extended the inference for every combination of discretized input variables c and ce. For example,

for c = –0.2 and ce = 0, v is –0.4

for c = –0.4 and ce = 0.4, v is 0.2

The corresponding lookup table is given in Table 10.8. Now we can use this lookup table in order to save the inference time and defuzzification time.

10.8 Fuzzy Expert Systems

An expert system is a program which contains human expert's knowledge and gives answers to the user's query by using an inference method. The knowledge is often stored in the form of rule base, and the most popular form is that of "if-then".

A fuzzy expert system is an expert system which can deal uncertain and fuzzy information. In our real world, a human expert has his knowledge in the form of linguistic terms. Therefore it is natural to represent the knowledge by fuzzy rules and thus to use fuzzy inference methods.

Table 10.8. Lookup table

e \ ce	−1.0	−0.8	−0.6	−0.4	−0.2	0	0.2	0.4	0.6	0.8	1.0
−1.0	−1.0	−1.0	−1.0	−1.0	−1.0	−0.8	−0.8	−0.8	−0.8	−0.6	−0.6
−0.8	−0.8	−0.8	−0.8	−0.8	−0.8	−0.8	−0.6	−0.6	−0.6	−0.6	−0.6
−0.6	−0.6	−0.6	−0.6	−0.6	−0.4	−0.4	−0.4	−0.4	−0.4	−0.2	−0.2
−0.4	−0.2	0	0	0	0	0	0	0.2	0.2	0.2	0.2
−0.2	−0.4	−0.4	−0.4	−0.4	−0.4	−0.4	0	0	0	0	0
0	−0.2	−0.2	−0.2	−0.2	−0.2	−0.2	−0.2	0	0	0	0
0.2	0.2	0.2	0.2	0.2	0.2	0.2	0.2	0.2	0.2	0.4	0.4
0.4	0.4	0.4	0.4	0.4	0.4	0.4	0.6	0.6	0.6	0.6	0.6
0.6	0.6	0.6	0.4	0.4	0.4	0.4	0.8	0.8	0.8	0.8	0.8
0.8	0.6	0.6	0.6	0.6	0.6	0.6	0.8	0.8	0.8	0.8	0.8
1.0	0.6	0.6	0.6	0.6	0.6	0.6	0.6	0.6	0.8	0.8	0.8

The structure of a fuzzy expert system is similar to that of the fuzzy logic controller. It's configuration is shown in (Fig 10.16.) As in the fuzzy logic controller, there can be fuzzification interface, knowledge base, and inference engine (decision making logic). Instead of the defuzzification module, there is the linguistic approximation module.

10.8.1 Fuzzification Interface

This module deals user's request, and thus we have to determine the fuzzification strategy. If we want to make the fuzzy expert system receive linguistic terms, this module has to have an ability to handle such fuzzy information. The fuzzification strategy, if necessary, is similar to that of the fuzzy logic controller.

Contrary to the fuzzy logic controller, it is not needed to consider the discretization or normalization. But the fuzzy partition and assigning fuzzy linguistic terms to each subregion are necessary.

The expert's knowledge may be represented in the form of "if-then" by using fuzzy linguistic terms. Each rule can have its certainty factor which represents the certainty level of the rule. This certainty factor is used in the aggregation of the results from each rule.

10.8.3 Inference Engine (Decision Making Logic)

The fuzzy expert systems can use the inference methods of the fuzzy logic controller. The system does not deal with a machine or process, and thus it is difficult to have a fuzzy set with monotonic membership function in the consequent part of a rule. Therefore especially, Mamdani method and Larsen method are often used.

10.8.4 Linguistic Approximation

As we stated before, a fuzzy expert system does not control a machine nor a process, and thus, in general, the defuzzification is not necessary. Instead of the defuzzification module, sometimes we need a linguistic approximation module.

This module finds a linguistic term which is closest to the obtained fuzzy set. To do it, we may use a measuring technique of distance between fuzzy sets.

10.8.5 Scheduler

This module controls all the processes in the fuzzy expert system. It determines the rules to be executed and sequence of their executions. It may also provide an explanation function for the result. For example, it can show the reason how the result was obtained.

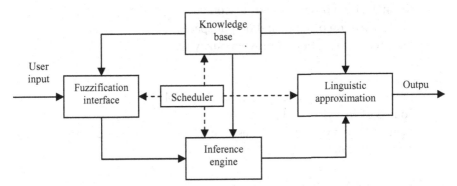

Fig. 10.16. Configuration of fuzzy expert system

[SUMMARY]

☐ The fuzzy logic controller (FLC) is good when
  – the control process is too complex
  – the information is qualitative

☐ Advantage of the fuzzy logic controller
  – Parallel and distributed control
  – Linguistic control
  – Robust control

☐ Components of the FLC
  – Fuzzification interface
  – Knowledge base
  – Decision-making logic
  – Defuzzification interface

☐ Fuzzification interface components
  – Scale mapping of input data
  – Strategy of noise
  – Selection of fuzzification function

☐ Data base components
  – Discretization of universe of discourse
  – Normalization of universe of discourse
  – Fuzzy partition of input and output spaces
  – Membership function of primary fuzzy set

☐ Rule base
  – Choice of state variables and control variables
  – Source of fuzzy control rules
  – Type of fuzzy control rules: state evaluation rules and object evaluation rules.

☐ Decision making logic
  – Mamdani inference method
  – Larsen inference method
  – Tsukamoto inference method
  – TSK inference method

☐ Defuzzification methods
  – Mean of maximum method (MOM)
  – Center of area method (COA)
  – Bisector of area (BOA)

☐ Lookup table
  – Disadvantage of the FLC is the high time complexity.
  – Saving of computation time
  – Direct relationship between input variable and control output actions

☐ Design procedure of the FLC
  – Determination of variables
  – Determination of inference method
  – Determination of fuzzification method
  – Discretization and normalization of variables
  – Partition of space
  – Determination of fuzzy sets
  – Construction of fuzzy rule base
  – Determination of defuzzification strategy
  – Test and tuning
  – Lookup table

[EXERCISES]

10.1 Describe advantage of the fuzzy logic controller (FLC).

10.2 In which case the FLC is superior to the conventional control algorithm?

10.3 Explain the followings:
   a) State variable
   b) Control output
   c) Control input
   d) Control variable

10.4 Explain the following components in the FLC.
   a) Fuzzification interface
   b) Knowledge base
   c) Data base
   d) Decision-making logic
   e) Defuzzification interface

10.5 Explain the three main issues in the fuzzification interface component

10.6 Explain the three main issues in the data base.

10.7 What is the difference between the discretization and normalization of universe of discourse.

10.8 Explain the relationship between the fuzzy partition of input variable and the number of fuzzy rules.

10.9 Why and how membership functions can be defined in a table when the universe of discourse is discretized?

10.10 What is the objective of the normalization of universe of discourse?

10.11 What are the main criteria to determine the state variables and control variables?

10.12 Explain the two-types of control rules:
   a) State evaluation fuzzy control rules
      R_i: if x is A_i, … and y is B_i then z is C_i
   b) Object evaluation fuzzy control rules

R_i: if (z is C_i → (x is A_i and y is B_i)) then z is C_i

10.13 Explain the four inference methods:

  a) Mamdani method
  b) Larsen method
  c) Tsukamoto method
  d) TSK method

10.14 What is the property of the membership function in consequent part in Tsukamoto method?

10.15 Explain the three defuzzification methods:

  a) Mean of maximum method
  b) Center of area method
  c) Bisector of area

10.16 What is the lookup table?   Why is it often used?

10.17 Show the design procedure of the FLC.

Chapter11. FUSION OF FUZZY SYSTEM AND NE-URAL NETWORKS

In this chapter, we study the fusion of fuzzy systems and neural networks. The two methods are complementary. The neural networks can learn from data while the fuzzy systems can not; the fuzzy systems are easy to comprehend because they use linguistic terms but the neural networks are not. Many researches have been devoted to the fusion of them in order to take their advantages.

11.1 Neural Networks

11.1.1 Basic Concepts of Neural Networks

Neural networks(NN) are a computational model of the operation of human brain. A neural network is composed of a number of nodes connected by links. Each link has a numeric weight associated with it. Weights are the primary means of long-term storage in neural networks. One of the major features of the neural network is its learning capability. They can adjust the weights to improve the performance for a given task. Learning usually takes place by updating the weights.

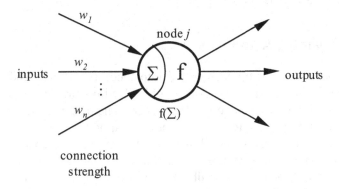

Fig. 11.1. Node of neural networks

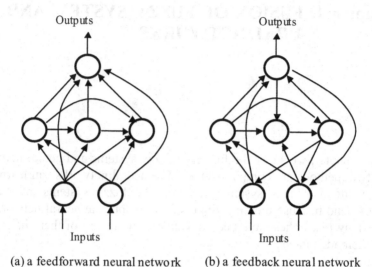

(a) a feedforward neural network   (b) a feedback neural network

Fig. 11.2. Neural Networks

A node has a set of input links from other nodes, a set of output links, and a nonlinear function called activation function. The output of a node is generated from the inputs, the weight of links, and activation function(Fig 11.1.). So, the output of node j are as follows:

$$o_j = f(net_j)$$

where $net_j = \sum_i w_{ij} o_i$, o_i is the output of node i and an input to node j, w_{ji} is the weight of the link between nodes j and i, and f is the activation function. Any nonlinear functions can be used as the activation function, but sigmoidal function is often used.

11.1.2 Learning Algorithms

Neural networks can be categorized into feedforward and feedback. The feedforward neural networks have only feedforward links, i.e. neural networks which do not have feedback cycle. The output of a node will not directly or indirectly be used as an input of that node.

To the contrary, the feedback neural networks have feedback cycle. The output of a node may be used as an input of a node. (Figs 11.2(a), (b)) show the feedforward and feedback neural networks, respectively. In the case of the feedback neural network, there is no guarantee that the networks become stable because of the feedback cycle. Some of them converge to a stable point, some may have limit-cycle, or become chaotic or divergent. These are common characteristics of non-linear systems whi-

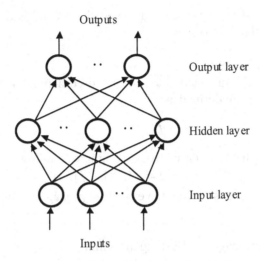

Outputs

Output layer

Hidden layer

Input layer

Inputs

Fig. 11.3. Multilayer perceptrons

ch have feedback.

There are two types of learning algorithms in the neural networks. The first type is supervised learning. It uses a set of training data which consist of pairs of input and output. During learning, the weights of a neural network are changed so that the input-output mapping becomes more and more close to the training data.

The second type is unsupervised learning. While the supervised learning presents target answers for each input to a neural network, the unsupervised learning has the target answers. It adjusts the weights of a neural network in response to only input patterns without the target answers. In the unsupervised learning, the network usually classifies the input patterns into similarity categories.

11.1.3 Multilayer Perceptrons and Error Backpropagation Learning

Among the neural networks and learning algorithms, multiplayer perceptron network and its learning algorithm are widely used. This learning algorithm is called error backpropagation method. Multilayer perceptrons are layered feedforward neural networks. They consist of several layers. A layer is a set of nodes which do not have inter-connection links, i.e. the nodes in the same layer are not connected to each other. One more characteristic is that the nodes in a layer are connected to only the nodes in the neighboring layer. (Fig 11.3) shows a three layer perceptron network. The first layer is the input layer, the second is the hidden layer and the third is the output layer.

(1)   Activation function
In this kind of neural networks, mostly used activation function is the sigmoidal function defined as:

$$f(x) = \frac{1}{1 + e^{-x}}$$

The reason is that their derivatives are simple as follows:
$$f'(x) = f(x)(1 - f(x))$$

If the derivative is simple, the error backpropagation algorithm can be simple.

(2)   Error backpropagation learning algorithm
The error backpropagation algorithm iteratively changes the weights of links little by little. The weights are changed so that the error is minimized. The error is defined as:

$$E = 0.5 \sum_k (t_k - o_k)^2$$

where t_k is the expected output and o_k is the actual outputs for node k in the output layer. In order to decide whether to increase or decrease weight w_{ji}, the algorithm evaluates the derivative of E with respect to w_{ji}, and then changes w_{ji} to the direction opposite to the derivation. That is,

$$w_{ji} := w_{ji} + \Delta w_{ji}$$

where

$$\Delta w_{ji} = -\eta \frac{\partial E}{\partial w_{ji}},$$

η is a learning rate and w_{ji} is the weight of the connection between node j and node i. Value η is a pre-determined small real number.

(3)   Convergence problem
Thus, if all weights of a neural networks are iteratively modified little by little in this manner, E will decrease and at last converge to a stable point. However it does not guarantee to converge to the global optimum because the method uses only derivations. So, in order to overcome this defect, some variations have been developed such as adding momentum term.

One more question on this algorithms is about magnitude of η. If η is a large value, the learning speed may be fast, but this may cause large oscillations. Small value will not cause the large oscillations, but the learning speed is slow. Choosing the value of η depends on applications.

(4)   Evaluation of change of weight

In order to evaluate Δw_{ji}, let us define $\delta_j = -\dfrac{\partial E}{\partial net_j}$, then

$$\frac{\partial E}{\partial w_{ji}} = \frac{\partial E}{\partial net_j} \times \frac{\partial net_j}{\partial w_{ji}} = -\delta_j o_i$$

because

$$\frac{\partial net_j}{\partial w_{ji}} = o_i .$$

thus,

$$\Delta w_{ji} = \eta \delta_j o_i .$$

What remains to evaluate Δw_{ji} is to evaluate δ_j.

If node j is an output node,

$$\delta_j = -\frac{\partial E}{\partial net_j} = -\frac{\partial E}{\partial o_j} \times \frac{\partial o_j}{\partial net_j} = (t_j - o_j)f'(net_j)$$

because

$$\frac{\partial E}{\partial o_j} = -(t_j - o_j)$$

$$\frac{\partial o_j}{\partial net_j} = f'(net_j)$$

If node j is a hidden node,

$$\delta_j = \frac{\partial E}{\partial o_j} \times \frac{\partial o_j}{\partial net_j} = -\left(\sum_k \frac{\partial E}{\partial net_k} \times \frac{\partial net_k}{\partial o_j}\right) \times \frac{\partial o_j}{\partial net_j} = \left(\sum_k \delta_k w_{kj}\right)f'(net_j)$$

because

$$\frac{\partial E}{\partial net_k} = \delta_k \quad \text{by definition}$$

$$\frac{\partial net_k}{\partial o_j} = w_{kj}$$

where node k is in the next layer, i.e. o_j is one of inputs of node k. The δ_js of the hidden nodes are evaluated from the summation of δ_ks of the neighboring layers.

If the sigmoidal function is used as the activation functions, it's derivative is

$$f'(net_j) = f(net_j)(1 - f(net_j)) = o_j(1 - o_j)$$

then the value of δ_j becomes for the output nodes

$$\delta_j = (t_j - o_j)o_j(1 - o_j)$$

and for the hidden nodes

$$\delta_j = o_j(1 - o_j)\sum_k (\delta_k w_{kj}).$$

11.2 Fusion with Neural Networks

Neural networks and fuzzy systems are two complementary technologies. Neural networks can learn from data, but the knowledge represented by the neural networks is difficult to understand. In contrast, fuzzy systems are easy to comprehend because they use linguistic terms and if-then rules, but it does not have learning algorithms. So, many researches have been devoted to fusion of them. The researches on fusion of neural networks and fuzzy systems can be classified into four categories:

(1) Modifying fuzzy systems with supervised neural network learning,
(2) Building neural networks using fuzzy systems,
(3) Making membership functions with neural networks,
(4) Concatenating neural networks and fuzzy systems.

11.2.1 Modifying Fuzzy Systems with Supervised Neural Network Learning

Research in this category represents fuzzy systems with neural networks. These systems are called neuro fuzzy systems, and the neural networks are used to improve the performance of fuzzy systems. Neuro fuzzy systems have characteristics of both neural networks and fuzzy systems; they have learning capability like neural networks and can perform inference like fuzzy systems.

In the ordinary neural networks, nodes have the same functionality and are fully connected to the nodes in the neighboring layers. But in a neuro fuzzy system, nodes have different functionalities and are not fully connected to the nodes in the neighboring layers. This differences come from the fact that the nodes and links in a neuro fuzzy system usually correspond to a specific component in a fuzzy system. That is, some nodes represent the linguistic terms of input variables, some nodes are for those of output variables, and some nodes and links are used for representing

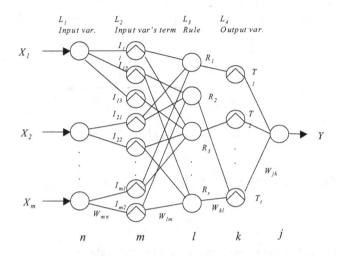

Fig. 11.4. Structure of the neuro fuzzy system proposed by Kwak

fuzzy rules.

One merit of the neuro fuzzy systems over the ordinary neural networks is the easiness of adding the expert knowledge before learning. The neuro fuzzy systems are usually built from the given fuzzy systems which are based on the expert or prior knowledge. Thus, the neuro fuzzy systems can embed the knowledge at the beginning. For this reason, convergence to local minimum may not be so much serious as that in the ordinary neural networks.

For example, the neuro fuzzy system proposed by Kwak, Lee and Lee-Kwang consists of five layers as shown in (Fig 11.4.). In the followings, the function of a node f is presented. Function f_j^k represents the f of node j of layer k.

(1) First layer of the network(Inputs)
   The nodes in the first layer takes inputs and just pass them to the second layer.

$$f_j^1(x) = x$$

(2) Second layer of the network(Input linguistic terms)
   A node in this layer represents a linguistic term of an input variable. It has parameters which represent the membership function of linguistic term. For example, in (Fig 11.4.). the input variable X1 is connected to three nodes in the second layer and X2 is connected to two nodes. It means that there are three linguistic terms defined on X1 and two linguistic terms on X2. The node in the second layer outputs the members degree

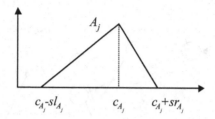

Fig. 11.5. A triangular fuzzy number

of input. For example, if a node represents a fuzzy set A, its output is $\mu_A(x)$.

$$f_j^2(x) = \mu_{A_j}(x)$$

where A_j is the fuzzy set represented by the node.

In this fuzzy neural network, only triangular fuzzy numbers are used to represent linguistic terms. A triangular fuzzy number A_j is defined with three parameters: the center c_{A_j}, the length of the left point from the center sl_{A_j}, and the length of the right point from the center sr_{A_j} as shown in(Fig 11.5.). So, the membership function of A_j is as follows:

$$\mu_{A_j}(x_i) = \begin{cases} 1 - \dfrac{x_i - c_{A_j}}{sr_{A_j}} & x_i \in (c_{A_j}, c_{A_j} + sr_{A_j}] \\[2mm] 1 + \dfrac{x_i - c_{A_j}}{sl_{A_j}} & x_i \in [c_{A_j} - sl_{A_j}, c_{A_j}] \\[2mm] 0 & otherwise \end{cases}$$

(3) Third layer of the network(Antecedent parts)

This layer corresponds to the antecedent parts of fuzzy rules. For example, in Fig. 11. the inputs of R_1 are the outputs of I_{11}, I_{21} and I_{m1}. It represents the antecedent part of the rule such as :

if X_1 is I_{11} and X_2 is I_{21} and $\cdots$ and X_m is I_{m1} then "

The output of the node is the matching degree of given inputs to the antecedent part. When evaluating the matching degrees, the minimum or product operators can be used. The connection weights between the second and third layers are fixed to 1.0.

$$f_i^3(x_1, x_2, \cdots, x_p) = \begin{cases} \min_{i=1}^p(x_i) & \text{if minimum used} \\ \prod_{i=1}^p(x_i) & \text{if product used} \end{cases}$$

(4) Fourth layer of the network(Consequent parts)

This layer represents the consequent parts of fuzzy rules. Like in the second layer, a node in this layer represents a linguistic term of output variable. For example, node T_t has two inputs from R_2 and R_r. It represents two rules whose consequent part is T_t:

"if the antecedent part is R_2 then Y is T_t"

and "if the antecedent part is R_r then Y is T_t"

The output of the node is the maximum matching degree of an input to the rules which are represented by the node. For example, the output of the node T_t is the maximum output of nodes R_2 and R_r. The weights between the third and fourth layers are used as the importance degree of rules, or fixed to 1.00.

$$f_j^4(x_1, x_2, \cdots, x_q) = \max_{i=1}^q\{w_{ji}x_i\}$$

where w_{ji} is the weight between node j in the fourth layer and node i in the third.

(5) Fifth layer of the network(Defuzzification)

A node in this layer gathers the outputs of all rules and defuzzifies them. A defuzzification method similar to the center of gravity method is used. The weight of links between it and the fourth layer is 1.00.

$$f_j^5(x_1, x_2, \cdots, x_t) = \frac{\sum_i^t Centroid(B_i, x_i)Area(B_i, x_i)}{\sum_i^t Area(B_i, x_i)}$$

where B_i is the fuzzy set represented by the node j in the fourth layer and x_i is the output of the node. If the output variable y is quantized with level n, $Centroid(B_i, x_i)$ and $Area(B_i, x_i)$ are defined as follows:

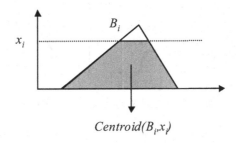

Centroid(B$_i$,x$_i$)

Fig. 11.6. Definition of *Centroid(B_i, x_i)*

$$Area(B_i, x_i) = \sum_i^n \min(\mu_{B_i}(y_i), x_i)$$

$$Centroid(B_i, x_i) = \frac{\sum_i^n y_i \cdot \min(\mu_{B_i}(y_i), x_i)}{\sum_i^n \min(\mu_{B_i}(y_i), x_i)} = \frac{C^u(B_i, x_i)}{Area(B_i, x_i)}$$

where $C^u(B_i, x_i) = \sum_i^n y_i \cdot \min(\mu_{B_i}(y_i), x_i)$. That is, $Centroid(B_i, x_i)$ is the centroid of the grey area in (Fig 11.6.). Thus, the output of this node can be rewritten as follows:

$$f_j^5(x_1, x_2, \cdots, x_t) = \frac{\sum_i^t C^u(B_i, x_i)}{\sum_i^t Area(B_i, x_i)}$$

(Fig 11.7.) shows the difference between the Kwak's defuzzification method and the center of gravity method. If we have the output result of fuzzy inference as shown (Fig 11.7.(a)), the output of the center of gravity method is

$$y_0 = \frac{\sum_i Centroid(B_i^c) \cdot Area(B_i^c)}{\sum_i^t Area(B_i^c)},$$

and the output of the Kwak's method is

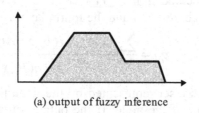

(a) output of fuzzy inference

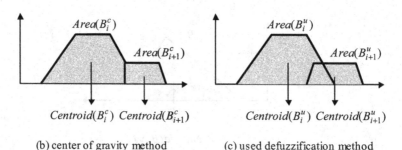

$Area(B_i^c)$ $Area(B_{i+1}^c)$

$Centroid(B_i^c)$ $Centroid(B_{i+1}^c)$

(b) center of gravity method

$Area(B_i^u)$ $Area(B_{i+1}^u)$

$Centroid(B_i^u)$ $Centroid(B_{i+1}^u)$

(c) used defuzzification method

Fig. 11.7. Defuzzification methods: the center of gravity method and the methods used by Kwak

$$y_0 = \frac{\sum_i Centroid(B_i^u) \cdot Area(B_i^u)}{\sum_i Area(B_i^u)}.$$

The difference is the how many times the overlapped areas are evaluated. The center of gravity method evaluates those areas only one time, but the Kwak's method may evaluate the overlapped area several times. One merit of the Kwak's method is that the defuzzification may be done with simpler operations than the center of gravity.

(6) Learning algorithm
The learning algorithm of this model is based on the error backpropagation. During the learning process, the weights between the third and fourth layers, and the parameters representing membership functions in the nodes of the second and fourth layers are modified based on the error backpropagation method.

11.2.2 Building Neural Networks using Fuzzy Systems

In the neuro fuzzy systems, neural networks were used to improve the performance of fuzzy systems. But the methods in this category use fuzzy system or fuzzy-rule structure to design neural networks. This model is a kind of divide and conquer approaches. Instead of training a neural network for the whole given input-output data, this model builds several networks:
(1) Builds a fuzzy classifier which clusters the given input-output data into several classes,
(2) Builds a neural network per class,

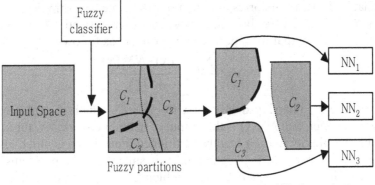

Fig. 11.9. Building neural networks using fuzzy systems

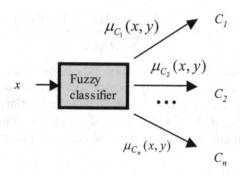

Fig. 11.9. A fuzzy classifier

(3) Trains the neural networks with the input-output data in the corresponding class.

(Fig.11.8.) shows the schematic diagram of this model. A fuzzy classifier divides the given data into several classes whose boundary is fuzzy. For building fuzzy classifier, many kinds of methods can be used such as fuzzy c-means, neural networks, fuzzy rules, and so on. (Fig 11.9.) shows a fuzzy classifier of which outputs are the membership degrees of input to the classes. Thus, we can say that this model has the following fuzzy rules:

$$\textbf{if } x \textbf{ is } C_1, \textbf{ then } y = NN_1(x)$$
$$\textbf{if } x \textbf{ is } C_2, \textbf{ then } y = NN_2(x)$$
$$\cdots$$
$$\textbf{if } x \textbf{ is } C_n, \textbf{ then } y = NN_n(x)$$

That is, a class can correspond to a rule. The membership degree of x to class C_i is calculated by the fuzzy classifier instead of membership functions. The output for input x can be evaluate as follows:

$$y = \frac{\sum_i \mu_{C_i}(x) \cdot NN_i(x)}{\sum_i \mu_{C_i}(x)}.$$

$C_i(x)$ is the membership degree of x to class C_i, which is evaluated by a fuzzy classifier, and $NN_i(x)$ is the output of neural network NN_i for input x.

The complexity in each fuzzily partitioned input space is much less than that of whole given task. This approach is suitable to the problems which can be divided into small sub-problems. For example, developing recognizer of characters of multiple fonts can be achieved by developing several recognizers of each font. If there is a font classifier, we can classify characters into each font group and then recognize the characters in each font group with the recognizer specialized to the corresponding font.

With this method, we can use priori knowledge of the task to reduce the complexity and increase the performance of a system. The fuzzy rules describe the priori knowledge of the given task, which can be obtained by analyzing training data.

An example of this model was proposed by Takagi and Hayashi. They used a neural network to build a fuzzy classifier, and proposed a method which could reduce the number of rules or redundant input variables. Their method is briefly summarized:

(1) Data preparation

The input/output data $(\mathbf{x}_i, y_i)$ are divided into the training set and the testing set. The numbers of the training and the testing set are n_{tr} and n_{ts}, respectively. The training set is used for building a neural network and the testing set is used for reducing the number of input variables. Let the number of inputs be m, i.e. $\mathbf{x}_i = (x_1, x_2, \cdots, x_m)$.

(2) Crisp partition of input space

The data in the training set are grouped into r classes, C^s, $s=1,2, \ldots, r$, by a crisp clustering method. In the training set, data belonging to C^s are denoted by $(\mathbf{x}_i^s, y_i^s)$, where $i=1,2,\cdots,n_{tr}^s$, and n_{tr}^s is the number of data belonging to C^s in the training set.

(3) Development of fuzzy classifier

This step builds a fuzzy classifier by using a neural network. The classifier will be denoted by NN_{mem} because it is a neural network which evaluates the membership degree of an input to each class. For training NN_{mem}, new data $(\mathbf{x}_i, \mu_i^s)$ are generated as follows:

$$\mu_i^s = \begin{cases} 1 & \mathbf{x}_i \in C^s \\ 0 & \mathbf{x}_i \notin C^s, \end{cases} \quad i=1,\cdots,n_{tr}^s; s=1,\cdots,r$$

Here, μ_i^s is the membership degree of $\mathbf{x}_i$ to C^s. NN_{mem} will be conducted so that these $\mu_i^s, i=1,\cdots,r$ can be inferred from the input $\mathbf{x}_i$. Thus, the NN_{mem} becomes capable of inferring the membership degree of an input to each class. $NN_{mem}(\mathbf{x})$ will denote the output of NN_{mem} for $\mathbf{x}$.

Though the training data for NN_{mem} have a sharp boundary between classes, NN_{mem} will change these sharp boundaries into fuzzy ones. Usually neural networks perform generalization of given data, and thus they generate smooth boundaries.

(4) Development of neural networks

This step is for building NN_s for C^s. The data $(\mathbf{x}_i^s, y_i^s), i = 1, 2, \cdots, n_i^s$, are used for training NN_s. This NN_s approximates the given data belonging to C^s. $NN_s(\mathbf{x})$ will denote the output of NN_s for $\mathbf{x}$. If this step is finished, the neural networks built in fuzzy-rule structure are obtained.

(5) Reduction of input variables

This step is for reducing the number of used input variables. The first action of this step is the evaluation of NN_s. In order to evaluate NN_s, the data in test set, $(\mathbf{x}_i, y_i), i = 1, 2, \cdots, n_{ts}$, is applied to NN_s, and the sum of error E_m^s is obtained as follows (m is the number of inputs):

$$E_m^s = \sum_{i=1}^{n_{ts}} \{y_i - NN_s(\mathbf{x}_i) \cdot NN_{mem}(\mathbf{x}_i)\}^2$$

Then, among the m input variables, an input variable x^p is arbitrarily eliminated, and NN_s is trained by the data without x^p in the training set as in the step 4. To evaluate the performance of the newly built neural network $NN_s^p(\mathbf{x}_i)$, E_{m-1}^{sp} is evaluated with the data (including x^p) in the test set as follows:

$$E_{m-1}^{sp} = \sum_{i=1}^{n_{ts}} \{y_i - NN_s^p(\mathbf{x}_i) \cdot NN_{mem}(\mathbf{x})\}^2$$

By comparing E_m^s and E_{m-1}^{sp}, if

$$E_m^s > E_{m-1}^{sp},$$

the significance of the network trained without x^p can be considered minimal, so x^p can be discarded. The same operations would hold for the remaining $m-1$ input variables.

(Fig 11.10.) shows the fuzzy-rule-structured neural network proposed by Takagi and Hayashi. The extended method was proposed in Takagi, Suzuki, Kouda and Kojima.

11.2.3 Making Membership Functions with Neural Networks

This approach is very similar to the previous one. In the previous approach, fuzzy-rule structure was used to build neural network. But in this approach, a neural network is used to generate compact fuzzy rules.

(1) Fuzzy partition of data

An important element of fuzzy systems is the fuzzy partition of input space. So, if there are k inputs, the fuzzy rules generate k-dimensional fuzzy hypercubes in the input space. Even though we can easily underst-

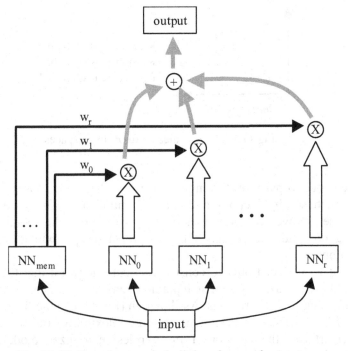

Fig. 11.10. Neural networks built in a fuzzy-rule structure

and these fuzzy hypercubes, it is not easy to get a flexible partition for nonlinear hypersurfaces. Since neural networks are proper to approximate nonlinear functions, they can be applied to construct a fuzzy partition.

The idea of this model is to build a neural network whose outputs are degrees that an input belongs to each class. These degrees can be considered as the membership degrees to each class. That is, the neural network takes the role of membership functions. Thus the membership functions of this model can be non-linear and multi-dimensional unlike the conventional fuzzy systems.

(2) Partition in a hyperspace

For example, we have two input variables x and y, and we know that the input space can be partitioned into three fuzzy classes: C_1, C_2 and C_3 as shown in Fig. 11.11. The gray boundary means that it is not clear so the points on the boundary could belong to all classes adjacent to the boundary.

Let us suppose that we can easily make fuzzy rules as long as we can partition the input space into C_1, C_2 and C_3. In this case, if we use fuzzy

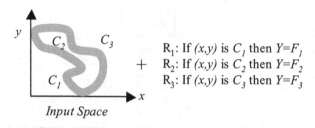

Fig 11.11. Fuzzy rules on fuzzy hyperspaces

hypercubes to partition the input space into the three classes, we need lots of fuzzy hypercubes and thus the number of fuzzy rules may increase. So, we need other method for partitioning. The neural network is a good candidate because they are proper to approximate nonlinear functions.

That is, we need only to construct a neural network that takes input variables of a fuzzy system as input and generates the degrees the input data belong to fuzzy regions. As shown in (Fig 11.11.) the fuzzy regions are fuzzy hypersurfaces. Due to the flexibility of these fuzzy hypersurfaces, the number of fuzzy rules in a fuzzy model can be reduced.

(3) Example of the rolling mill control
This model has been used in the control of rolling mill. The purpose of the rolling mill is to make plates of iron, stainless, or aluminum by controlling 20 rolls. The controller will change system parameters if the

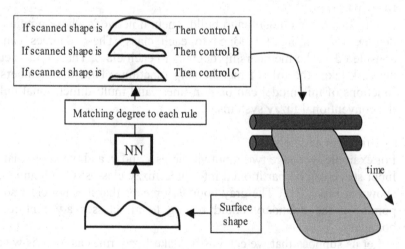

Fig. 11.12. Fuzzy rules with fuzzy hyperspace

surface of plate is not flat. How to change the parameters is determined by the produced surface shape of plates. To do this, first fuzzy rules are generated for only 20 standard template surface shapes. Then, a neural network is constructed which generates the similarity degrees to which an arbitrary surface shape belongs to each standard template shape.

The similarity degrees produced by the neural network are used as the matching degrees to the antecedent part of each rule. Since the antecedent parts of fuzzy control rules are standard template surface patterns, the output of the neural network corresponds to how much input surface pattern matches to each fuzzy rule. (Fig 11.12.) shows the structure of the system.

11.2.4 Evaluation of Fuzzy Systems with Neural Networks

When a fuzzy system is developed, we have to evaluate the system in order to tune it or know its performance. When it is not easy to apply the developed system to the real system, neural networks can be used as a simulator of the real system.

Jung, Im, and Lee-Kwang developed a fuzzy control algorithm for shape control in cold rolling of a steel work. They modeled strip shapes and constructed a fuzzy rule base producing control actions for each irregular strip shape. They identified parameters Λ_1, Λ_2, Λ_3 and Λ_4 representing the cross sectional shapes(Fig 11.13.), and by using them, fuzzy rules are obtained such as:

　　IF Λ_2 is *LPB* **THEN** ΔF_w is *LPB*, ΔF_i is *PB*,

　　IF Λ_2 is *PB*　**THEN** ΔF_w is *PB*, ΔF_i is *PM*,

where ΔF_w and ΔF_i are changes of the work forces applied to the rolls.

In order to evaluate the fuzzy rule base system before implement the system in the real field, they developed an emulator simulating the real cold rolling machine by using a neural network(Fig 11.14.). By using the neural network, they tuned the developed fuzzy system and then obtained desired results.

11.2.5 Concatenating Neural Networks and Fuzzy Systems

This category includes the methods equally using fuzzy systems and neural networks to improve system performance.

(1) Parallel combination

　　This combination is for correction of the output of a fuzzy system with the output of a neural network to increase the precision of the final system output. Fig. 11.(a) shows this combination. If a fuzzy system exists and an input-output data set is available, this model can be used to

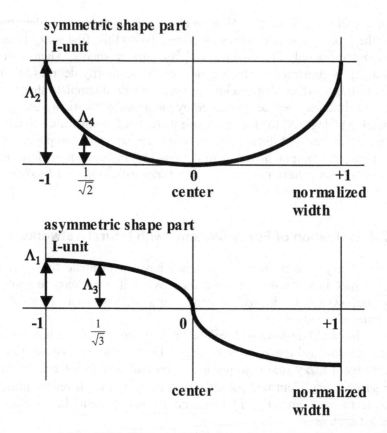

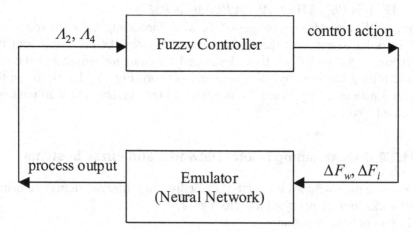

Fig. 11.13. Representation of the shape parameters

Fig. 11.14. Structure of the emulator

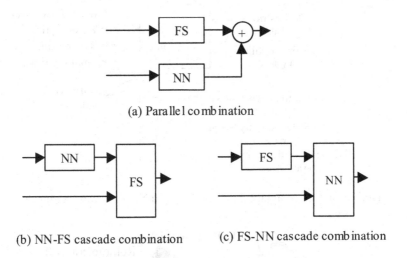

(a) Parallel combination

(b) NN-FS cascade combination   (c) FS-NN cascade combination

Fig. 11.15. Possible concatenations of NNs and FSs

improve the performance without modifying the existing fuzzy systems. For example, there is a fuzzy system working, but to improve its performance we need to incorporate more inputs to the fuzzy system. Usually designing a new fuzzy system is very costly. So, rather than redesigning the entire fuzzy systems to handle the extra inputs, a neural network is added that it manages the new input and corrects the output.

Examples of this method can be found in consumer products. The consumer product companies usually upgrade their products by adding some functionality. Consequently, the controller of the upgraded product usually requires more inputs to deal with such new functions while it performs almost the same functions as the controller of the old model.

In this case, modifying old one can be better than developing a totally new controller. Redesigning the fuzzy system is thus avoided, which causes a substantial saving in development time and cost, because redesigning the membership functions becomes more difficult as the number of inputs increases.

(Fig 11.16.) shows the schematic underlying a Japanese washing machine. The fuzzy system is a part of the first model. Later in order to improve the performance, more inputs are added and a neural network is built to correct the output with the new inputs. The additional input fed only to the neural network is electrical conductivity, which is used to determine washing time, rinsing time and spinning time. To train the neural network, the desired correction is used. This value is the difference between the desired output and what the fuzzy system outputs.

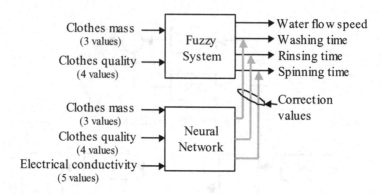

Fig. 11.16. Correcting mechanism of FS and NN for a washing machine

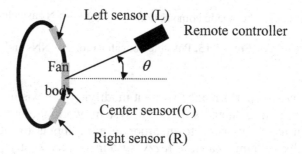

(a) the angle between remote controller and fan

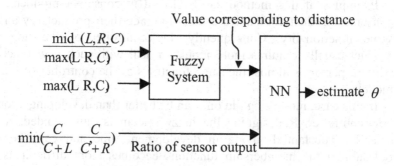

(b) cascade combination of NN and FS

Fig. 11.17. Cascade combination of FS and NN for a electric fan

(2) Cascade combination

The second is the cascade combination. It is a system where the output of the fuzzy system or neural network becomes the input of the other.

(Fig 11.15.(b)) shows the model where the output of a fuzzy system is inputted to a neural network and (Fig 11.15.(c)) shows the reverse model.

An example of (Fig 11.15.(c)) can be found in Japanese electric fan. One of the requirement of it is to rotate toward the user by detecting the signal of remote controller. For the detection of the signal, three sensors are equipped on the fan body as shown in (Fig 11.17.(a)). In order to estimate the angle from the center of fan to the user, θ, a neural network and a fuzzy system are used. The fuzzy system estimates the distance from the fan to the user by using the sensors. Then, the estimated distance is used as an input of the neural network which estimates θ. (Fig 11.17.(b)) shows the configuration of this application.

[SUMMARY]

☐ Neural Network(NN)
  – Nodes
  – Links
  – Weights
  – Activation function

☐ Characteristics of NN
  – Computational model of the brain
  – Learning capability
  – Learning by updating weights

☐ Structure of NN
  – Feedforward NN
  – Feedback NN

☐ Learning algorithms
  – Supervised learning
  – Unsupervised learning

☐ Multilayer perceptrons
  – Feedforward network
  – Error backpropagation learning algorithm
  – Convergence problem

☐ Fusion method of fuzzy systems and neural networks
  – Modifying fuzzy systems with NN
  – Building NN in fuzzy-rule structures
  – Making membership functions with NN
  – Concatenating NN and fuzzy systems

☐ Modifying fuzzy systems with NN
  – Learning capability
  – Inference capability

☐ Building NN in fuzzy-rule structure
  – Fuzzy classifier of data
  – NN per class
  – Learning of NN

☐ Making membership functions with NN
   – Fuzzy partition of data
   – Partition of a hyperspace

☐ Concatenating NN and fuzzy systems
   – Parallel combination
   – Cascade combination

[EXERCISE]

11.1 Why the fuzzy systems and neural networks are said to be complementary technologies?

11.2 Explain the structure of multilayer perceptrons and their learning algorithm.

11.3 What is the roles of the activation function and weights of links?

11.4 What is the mechanism of learning in the backpropagation method?

11.5 What is the convergence problem in the backpropagation learning algorithm?

11.6 Explain the properties of the fusion method "modifying fuzzy systems with supervised neural network learning".

11.7 In which case, the fusion "building NN in fuzzy-rule structure" is useful?

11.8 Explain concatenating method of NN and fuzzy systems.

Chapter 12. FUSION OF FUZZY SYSTEMS AND GENETIC ALGORITHMS

The fuzzy systems and genetic algorithms are complementary techniques because the fuzzy systems are easy to understand but the genetic algorithms are not; the genetic algorithms have an ability of learning while the fuzzy systems have no. In this chapter, some researches on the fusion of the two techniques are introduced.

12.1 Genetic Algorithms

12.1.1 General Structure of Genetic Algorithms

Genetic algorithms can be viewed as a general-purpose search method, an optimization method, or a learning mechanism. Their basic mechanisms are similar to Darwinian principles of biological evolution: reproduction and "survival of the fittest". Genetic algorithms maintain a set of candidate solutions. The set is called a population and candidate solutions are called individuals or chromosomes. Chromosomes are usually represented in binary strings of a fixed length. Due to the set of candidate solutions, the genetic algorithms are inherently parallel. The genetic algorithms have been shown to be an effective search technique on a wide range of difficult optimization problems.

(1) Operations in genetic algorithms
   A typical genetic algorithm performs a sequence of operations on a population as follows:

   1) Initialize a population of chromosomes (population size $= n$).
   2) Evaluate the fitness of each chromosome in the population.
   3) If the stop condition is satisfied, stop and return the best chromosome in the population
   4) Select $n/2$ pairs of chromosomes from the population. Chromosomes can be selected several times.
   5) Create new n chromosomes by mating the selected pairs by applying the crossover operator.

6) Apply the mutation operator to the new chromosomes.
7) Replace the old population with the new chromosomes.
8) Goto (2).

This cycle is largely divided into evaluation, selection and reproduction. Step 3 is for the evaluation, Step 4 for the selection, and Step 5 and 6 for the reproduction. This cycle terminates when an acceptable solution is found, when a convergence criterion is met, or when a predetermined limit on the number of iterations is reached.

(2) Encoding scheme
Encoding scheme is how to represent candidate solutions into binary strings. Since genetic algorithms manipulate binary strings, the candidate solutions are needed to be represented as binary strings. If the given problem is finding a point in a multi-dimensional space such as maximizing a function, then the candidate solutions can be simply encoded into binary strings.

For example, if only integers greater than 0 and less than 255 are used, a triplet of $(10,41,21)$ can be encoded as follows:

$$00001010\ 00101001\ 00001101$$

The first eight bits 00001010 are the binary representation of decimal 10, the next eight bits 00101001 are that of 41, and the last eight are that of 21.

However, if some structures such as trees or matrix are candidate solutions, we need to develop a way to encode them into binary strings. Since chromosomes represent the candidate solutions, the encoding scheme should be designed so that it can cover the possible solution space. Also, it should be designed so that a modification of chromosomes through genetic operators such as crossover and mutation, is easy and effective. Sometimes, the chromosomes are encoded into real value strings, or symbolic strings instead of binary strings.

12.1.2 Evolution of Genetic Algorithms

(1) Fitness function and evaluation
We need a function to evaluate the fitness of candidate solutions during the operations. The fitness of chromosomes are the driving force of genetic algorithms. It represents how much good solution a chromosome is. Thus, the fitness function should be designed so that it gives higher fitness values to better solutions. If there is a function to be optimized, it is usually used as a fitness function.

For example, if the maximum of $f(x)=-x^2+4x-1$ on $\{x \mid 0 \le x \le 2\}$ are to be found and the value x is encoded, then the function $f(x)$ can be directly used as the fitness function. For instance, the fitness value of a chromosome representing 1 is $f(1)=2$.

However, if we do not have such functions, we need to devise one which properly evaluates the quality of the solution. For example, in the case of building a decision tree with given data, we should design a fitness function based on the variables which can reflect the quality of a decision tree such as the number of nodes, the correctness of output, and so on.

(2) Selection

Selection is an operation which prepares reproductions. The selected chromosomes are called parents. For the selection, first the possibility for each chromosome to be selected is evaluated. This possibility largely depends on the fitness value; the higher fitness value, the higher selection possibility. The reason is that it is expected that better offspring can be generated from better parents. There are several selection methods.

1) Roulette wheel selection

The roulette wheel selection is a typical one. The selection probability of a chromosome is the ratio of its fitness value to the sum of those of all chromosomes. That is, this method gives the selection probability to individuals linearly proportional to their fitness values. For example, there are five chromosomes, I_1, I_2, I_3, I_4 and I_5, and their fitness values are 1, 4, 3, 6, and 2, respectively. The summation of fitness is 16. Thus, the selection probability of I_1 is $1/16=0.0625$, I_2 is $4/16=0.25$, I_3 is $3/16=0.1875$, I_4 is $6/16=0.375$ and I_5 is $2/16=0.125$. Each chromosome will be selected based on these probabilities.

2) Rank based selection

The other selection method is the rank-based selection. In the roulette wheel method, the selection probability is linearly proportional to the fitness values, but in rank-based selection, the selection probability is fixed according to the rank of the fitness. For example, if the selection probabilities are given as follows: $(0.3, 0.25, 0.2, 0.15, ...)$, the chromosome with the maximum fitness always has a selection probability of 0.3, the chromosome with the second largest has 0.25 and the others have the probability in this manner.

One merit of this method over the roulette wheel method is preventing fast convergence to a local maximum. If a few chromosomes have a very high fitness, and others have very small. Then, if the roulette wheel method is used, the chromosomes with high fitness also have high selection probabilities while the others have very low. Thus, the chromosomes with high probability are almost always selected and thus most offsprings are generated from them. It makes the variety of a population low, and thus all chromosomes are easy to stick to a local maximum.

(3) Crossover

Crossover operators produce two new chromosomes by exchanging information of the selected chromosomes. This operator is the most essential in genetic algorithms. The most typical crossover operator is the one-point crossover. The selected chromosomes are cut on the randomly chosen point, and the cut parts are exchanged. It is shown in (Fig 12.1(a)). An extension is the multi-points crossover in which several points are chosen. (Fig 12.1(b)) shows the two points crossover.

The crossover operations are not performed on every selected chromosome. Genetic algorithm decides, based on a given probability, whether it performs the crossover operation on the certain pair of chromosomes or not. It is called the crossover probability and given by users.

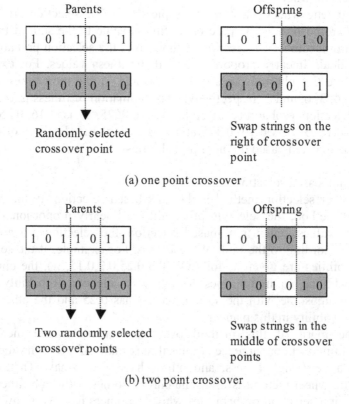

Fig. 12.1. Crossover operator

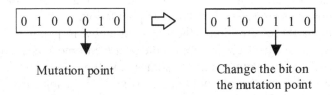

Mutation point   Change the bit on
                                        the mutation point

Fig. 12.2. Mutation operator

(4) Mutation

Mutation operators change some randomly selected bits of chromosomes. If the chromosomes are binary strings, then '0' are changed to '1', and '1' to '0'. It plays a secondary role after the crossover operator in genetic algorithms. The changing bits means making an offspring genetically different from its parents. Since the crossover operator mixes the information of only two parents, the information of offsprings may not be much different from that of its parents. Thus, applying only the crossover operator may make a population trapped in a local optimum.

In order to escape from a local optimum, a kind of jump operation is needed. So, by using the mutation operator, we can get some offsprings different from their parents. That is, the genetic algorithms try to jump to other place. However, if the mutation operator is often applied to chromosomes, then most of newly created chromosomes are randomly different from their parents, so that the searching process may loss its direction. Thus, the mutation probability of a bit should be very low.

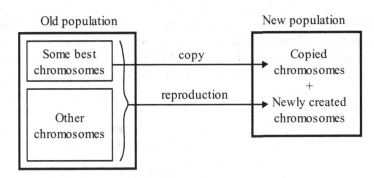

Fig. 12.3. Elitist strategy

(5) Replacement

A typical genetic algorithm totally replaces the old population with the newly created chromosomes, but it is not mandatory. There could be many variations. For example, after reproduction, the old and new populations are taken together, and among them the best n chromosomes are selected as the next population.

Among these variations, the elitist strategy is popular. The elitist strategy is an approach that copies the best k chromosomes into the next population. The other chromosomes of the new populations are reproduced from the old population. (Fig 12.3) graphically shows the elitist strategy. If the total replacement is performed, the chromosome of the best fitness in the new population may be worse than that in the old population. This is why the elitist strategy is useful.

12.2 Fusion with Genetic Algorithms

Like in the fuzzy systems and neural networks, the fuzzy systems and genetic algorithms can complement each other. Researches on their fusion of them can be classified into two categories:

(1) Identifying fuzzy systems with genetic algorithms

(2) Controlling parameters of genetic algorithms with fuzzy systems.

As mentioned, the fuzzy systems do not have learning algorithms, so the genetic algorithms can be used as a learning algorithm of the fuzzy systems. Genetic algorithms have some parameters to be set, so fuzzy rules can be used to change these parameters during the searching process.

12.2.1 Identifying Fuzzy Systems with Genetic Algorithms

Although fuzzy systems have been used to control a number of systems, the selection of acceptable fuzzy membership functions has been a subjective and time-consuming task. When we build a fuzzy system, we should determine the number of the linguistic terms of input and output

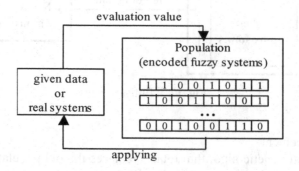

Fig. 12.4. Schematic diagram of identifying FSs with GAs

Although fuzzy systems have been used to control a number of systems, the selection of acceptable fuzzy membership functions has been a subjective and time-consuming task. When we build a fuzzy system, we should determine the number of the linguistic terms of input and output variables, their membership functions and the consequence parts of fuzzy rules.

The IF-THEN structure of fuzzy rules is easy to understand and to build with priori knowledge, but many parameters should be specified by experts. The identification of these parameters can be viewed as an optimization problem; finding parameters that optimize the performance of the model. Therefore, there have been many researches on applying the genetic algorithms to the identification of fuzzy systems. These researches encode the parameters of a fuzzy system into chromosomes, and these chromosomes are evolved to find parameters which make a fuzzy system fit to real systems or given data well.

(Fig 12.4.) illustrates the evolving process, and the researches can be categorized into groups:

(1) Tuning an existing fuzzy system

The researches in the first category modify the parameters of an existing fuzzy system. The usually tuned parameters are the membership functions and/or fuzzy rules. Tuning the membership functions with genetic algorithms are analogue to neuro fuzzy systems. In these researches, the membership functions are encoded into chromosomes and better membership functions are searched by genetic algorithms.

(Fig 12.5.) shows an example of encoded fuzzy sets into chromosomes. In this example, the center of triangular fuzzy set is fixed, and only the left and the right points of each fuzzy set are variable and thus encoded. To modify the fuzzy rules, their consequent parts are usually encoded. For example, there are four fuzzy rules:

$$\text{IF } X \text{ is } I_1 \text{ THEN } Y \text{ is } O_1$$
$$\text{IF } X \text{ is } I_2 \text{ THEN } Y \text{ is } O_2$$
$$\text{IF } X \text{ is } I_3 \text{ THEN } Y \text{ is } O_3$$
$$\text{IF } X \text{ is } I_4 \text{ THEN } Y \text{ is } O_4$$

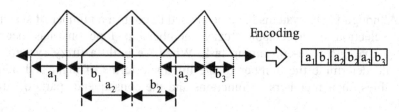

Fig. 12.5. Example of encoding fuzzy sets

   then, these are encoded as a string of linguistic terms like $O_1 O_2 O_3 O_4$.
The genetic operators will change the linguistic terms, but not their
membership functions. For example $O_1 O_2 O_3 O_4$ may be changed into
$O_1 O_3 O_4 O_1$ after genetic operations. This represents the following fuzzy
rules:

$$\text{IF } X \text{ is } I_1 \text{ THEN } Y \text{ is } O_1$$

$$\text{IF } X \text{ is } I_2 \text{ THEN } Y \text{ is } O_3$$

$$\text{IF } X \text{ is } I_3 \text{ THEN } Y \text{ is } O_4$$

$$\text{IF } X \text{ is } I_4 \text{ THEN } Y \text{ is } O_1$$

this approach is proper to rough tuning of fuzzy systems because
changing rules may affect the fuzzy system.

(2) Building a fuzzy system with genetic algorithm
   This method do not need an existing fuzzy system. This approach
determines all the parameters of a fuzzy system by genetic algorithms
without any priori knowledge. Thus, the chromosomes used in this
method usually include most of the parameters such as the number and
membership functions of linguistic terms. So, it is very important how
to effectively represent those parameters because a long chromosome
means a wide search space.
   If a search space is wide, we cannot expect a good optimization
result. So, most researches make restrictions; for example, some fix the
number of linguistic terms or restrict the shape and position of
membership functions.
   In a sense that this approach does not use any priori knowledge, it is
analogue to neural networks. One merit of this approach over neural
networks is that we can easily extract knowledge on systems. This
approach represents discovered knowledge with fuzzy rules and fuzzy
sets. Thus, we can easily understand and extract the knowledge. In the
case of neural networks, we cannot understand the knowledge
represented inside the networks.
1) Example

There is an example to build fuzzy systems by genetic algorithms. Lee and Lee-Kwang proposed a method which constructs a fuzzy system by using given input-output data. They assumed that no priori knowledge nor existing fuzzy system was given, and that only input-output data were available. Thus, all parameters of a fuzzy system needs to be identified.

If all fuzzy sets and rules are found by genetic algorithms, they need to be encoded into chromosomes. However, it would produce long bit strings. A long string leads to broadening the search space, and as the result, its searching performance may get bad effects. To cope with it, they encoded only fuzzy sets into chromosomes. When building a fuzzy system from a chromosome, fuzzy sets are decoded from the chromosome, and then fuzzy rules are generated based on the given data and the decoded fuzzy sets. They used the product operator for the evaluation of antecedent matching degree.

2)   Encoding scheme of input variables

To represent a fuzzy model by a binary string, an encoding scheme is needed. In this method, only the position of fuzzy sets of input variables are encoded which can be interpreted as a fuzzy partition. The encoding scheme is devised so that the number, the position and the size of fuzzy sets can evolve during the searching process.

In this model, only triangular fuzzy sets are used. For a triangular fuzzy set, the specification with three points is needed: two base points and one center point. To easily handle the varying number of fuzzy sets and their positions, an encoding scheme for input variables is used as shown in (Fig 12.6.) The input domain is discretized into L points. Only the center point of a fuzzy set is represented by one of the L points. The "1" in a bit string indicates the center point of a fuzzy set. So the numb-

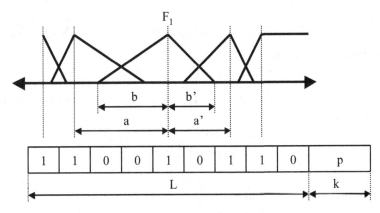

Fig. 12.6. Encoding scheme of the method proposed by Lee and Lee-Kwang

er of the "1"s is equal to the number of fuzzy sets, and L is it's maximum number in an input domain.

A chromosome contains another parameter. That is the overlapping coefficient p. It takes k bits, and we get the value of p by decoding them. It is also used to determine the position of fuzzy sets.

For example, the positions of fuzzy set F_1 in (Fig 12.6) is determined as follows. We can get the position of the center point from the location of the "1" in chromosomes. Then, what remains is the position determination of the two other base points. These points are located dependently on the center point of the neighboring fuzzy sets. The left point is located at $b = a \times p$ off from the center where a is the distance between the centers of the fuzzy set F_1 and the left neighboring fuzzy set. Similarly, the right point is at $b' = a' \times p$ plus to the center. Since $L + k$ bits are needed for an input variable, in the case of n inputs, the length of chromosomes becomes $n \times (L + k)$.

3) Determination of consequent parts

Since only the fuzzy partitions of each input space were encoded in the above, fuzzy rules should be generated now. To get the fuzzy rules from a chromosome, the basic form of fuzzy rules should be built from the cartesian product of the fuzzy partitions.

For the specification of consequent part, the given data and the least square estimation method are used. The detail procedure is explained in the following: Let us assume that there are only two input variables.

First, let's suppose we have n_r incomplete rules with undefined consequent parts, C_i, $i=1, \ldots, n_r$,

$$R_1 : \text{IF } x_1 \text{ is } A_{11} \text{ and } x_2 \text{ is } A_{12} \text{ THEN } y = C_1$$

$$R_2 : \text{IF } x_1 \text{ is } A_{21} \text{ and } x_2 \text{ is } A_{22} \text{ THEN } y = C_2$$

$$\vdots$$

$$R_{n_r} : \text{IF } x_1 \text{ is } A_{n_r 1} \text{ and } x_2 \text{ is } A_{n_r 2} \text{ THEN } y = C_{n_r}$$

and n_s sample data

$$(x_{11}, x_{12}, y_1)(x_{21}, x_{22}, y_2), \cdots, (x_{n_s 1}, x_{n_s 2}, y_{n_s})$$

For the ith sample (x_{i1}, x_{i2}, y_i), the output y_i' is obtained by the following way:

$$y_i' = \frac{\sum_{k=1}^{n_r} \mu_k(x_{i1}, x_{i2}) \cdot C_k}{\sum_{k=1}^{n_r} \mu_k(x_{i1}, x_{i2})}$$

then, the above equation is

$$y_i' = a_{1i} C_1 + a_{2i} C_2 + \cdots + a_{n_r i} C_{n_r}$$

where

$$a_{ji} = \frac{\mu_j(x_{i1}, x_{i2})}{\sum_{k=1}^{n_r} \mu_k(x_{i1}, x_{i2})}$$

for all samples, the outputs can be written like the followings.

$$y_1' = a_{11}C_1 + a_{21}C_2 + \cdots + a_{n_r 1}C_{n_r}$$

$$y_2' = a_{12}C_1 + a_{22}C_2 + \cdots + a_{n_r 2}C_{n_r}$$

$$\vdots$$

$$y_{n_s}' = a_{1n_s}C_1 + a_{2n_s}C_2 + \cdots + a_{n_r n_s}C_{n_r}$$

In order to obtain the generated output y_i' as close to y_i as possible, the least square method is used. That is, the sum of the squared error between the outputs and the given data should be minimized. The method gives the value of C_i, $i=1, \ldots, n_r$, such that the sum is minimized

$$\sum_{k=1}^{n_s} (y_k - y_k')^2 .$$

We can get those C_i, $i=1, \ldots, n_r$, from some matrix manipulations. Through this way, for a given fuzzy partition, we can get the fuzzy rules which produce the output as close as the given data.

4) Cost function

To evaluate the fuzzy model encoded in a chromosome, the number of fuzzy rules, the sum of error, the maximum of error and the number of sample data are considered. The error is defined as the absolute value of the difference between the output and expected value.

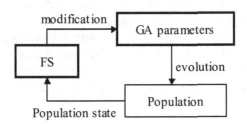

Fig. 12.7. Schematic diagram of controlling parameters of GAs with FSs

12.2.2 Controlling Parameters of Genetic Algorithms with Fuzzy Systems

Genetic algorithms need some parameters such as population size, and probabilities of crossover and mutation. These parameters are very important for the performance, and the interaction between them is known to be complex. Thus, there have been many researches on how these parameters effect on the performance and how to set them to improve the performance.

However, the parameters setting is often left to the user and never changed during evolutions. There are some researches on dynamic control of parameters during evolutions by using fuzzy systems. The basic idea is simple; a fuzzy system observes the states of population during evolutions and changes the parameters to improve the performance. That is, genetic algorithms use a fuzzy knowledge-based system to dynamically control the parameters, such as population size, crossover rates, and mutation rates. (Fig 12.7.)shows the schematic diagram of this method. For example, fuzzy rules for those systems can be described as follows:

- **IF** average fitness is high **THEN** population size should be increased.
- **IF** best fitness is not improved **THEN** mutation rate should be increased.

One question of this approach is how to obtain the knowledge to build the fuzzy rules. It can be solved in the ways; an expert on genetic algorithms can describe his/her own knowledge or an automatic fuzzy design technique can be applied.

(1) Example of DPGA

An example of this approach is Dynamic Parametric Genetic Algorithms (DPGA) proposed by Lee and Takagi. The DPGA has a fuzzy system which controls the parameters of genetic algorithms according to the population state. At first, the fuzzy systems of the DPGA was manually built from empirical knowledge, but it did not show good performance. Thus an automated method to build fuzzy systems was used, which is similar to the method presented in the previous section. The fuzzy system was built by using genetic algorithms.

(2) Encoding scheme

To build a fuzzy system which can control the parameters of genetic algorithms well, parameters of the fuzzy systems are encoded into chromosomes. The inputs of the fuzzy system are (average fitness)/(best fitness), (worst fitness)/(average fitness) and the change of fitness since last control action. The outputs are the changes of population size,

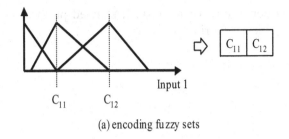

(a) encoding fuzzy sets

C_{11}	C_{12}	....	C_{32}	R_1	R_2	..	R_{81}	population size	crossover rate	mutation rate

(b) a chromosome representing a fuzzy system
C_{ij}: center point of the jth fuzzy set of Input I
R_i : the consequent part of ith rule(fuzzy set for output)

Fig. 12.8. Coding of a fuzzy set and a chromosome of the DPGA

crossover probability, and mutation probability. All the input and output variables are represented by three fuzzy sets as shown in (Fig 12.8.(a)).
To encode a fuzzy set, overlapped membership functions are used of which right and left end points are located at the center points of the right and left neighboring fuzzy sets, respectively. The center point of the left most set is fixed as shown in (Fig 12.8.(a)), so only two center points are variable and thus encoded. The total number of encoded parameters is 12 (two center points per input and output variables) + 81(rules).

In addition, initial values for population size, crossover and mutation rates are included in the coding as shown in (Fig 12.9.(b)).

(3) Evolution and evaluation
For the evolutions of chromosomes, some heuristics are added; the fuzzy system cannot change the population size by more than half of the current population size and should keep it between 10 and 160. The crossover and mutation rates are also restricted to change at most by half of their current value and are bounded by [0.2,1.0] and [0.0001, 1.0] respectively.
For the evaluation of chromosomes, DeJong's five functions are used. The fitness of a chromosome is evaluated based on how much DeJong's functions are optimized by the DPGA. The proposed DPGA has an initial population size of 13, crossover and mutation rates of 0.9 and 0.08 respectively. This DPGA may not show a good performance in other applications because it is optimized with respect to only DeJong's functions. In order to test the generality of the DPGA, it was applied to

an inverted pendulum control task. It showed performance as good as
the simple genetic algorithms.

[SUMMARY]

☐ Fuzzy systems
   – Easy to understand
   – No ability of learning

☐ Genetic algorithms
   – Not easy to understand
   – Ability of search

☐ Operation in genetic algorithms
   – Crossover
   – Mutation

☐ Issues in genetic algorithms
   – Encoding scheme
   – Selection
   – Crossover rate
   – Mutation rate
   – Fitness function and evaluation

☐ Fusion of fuzzy systems and genetic algorithms
   – Identifying fuzzy systems with genetic algorithms
     • tuning an existing fuzzy system
     • building a fuzzy system
   – Controlling parameters of genetic algorithms with fuzzy
     systems

☐ Issues in building a fuzzy system with genetic algorithms
   – Encoding scheme of input variables
   – Determination of consequent parts
   – Cost function

[EXERCISE]

12.1 Explain the advantages and disadvantages of the fuzzy system in comparing with the genetic algorithms.

12.2 Show the typical operation procedure of the genetic algorithms.

12.3 Explain the concepts and functions of the followings.
   a) Encoding scheme
   b) Crossover operation
   c) Mutation operation
   d) Fitness function
   e) Crossover rate
   f) Mutation rate

12.4 Explain two fusion methods of fuzzy systems and genetic algorithms: identifying fuzzy systems with genetic algorithms and controlling parameters of genetic algorithms with fuzzy systems.

BIBLIOGRAPHY

1.   Aburdene, M. F (1988) *Computer Simulation of Dynamic Systems*, WM. C. Brown Publishers, Dubuque
2.   Aleksander, I (1989) *Neural Computing Architectures*, MIT Press, Cambridge, Mass
3.   Aminzadeh, F. and Jamshidi, M (1994) Soft Computing - Fuzzy Logic, Neural Networkds and Distributed Artificial Intelligence, PTR Prentice Hall, Englewood Cliffs, N. J
4.   Angeline, P. J. and Kinnear, K. E. Jr (1996) *Advances in Genetic Programming*, MIT Press, Cambridge, Mass
5.   Antognetti, P. and Milutinovic, V (1992) *Neural Networks - Concepts, Applications and Implementations,* Vol 1, Prentice Hall, Englewood Cliffs, N. J
6.   Antognetti, P. and Milutinovic, V (1991) *Neural Networks - Concepts, Applications and Implementations,* Vol 2, Prentice Hall, Englewood Cliffs, N. J
7.   Antognetti, P. and Milutinovic, V (1991) *Neural Networks - Concepts, Applications and Implementations,* Vol 3, Prentice Hall, Englewood Cliffs, N. J
8.   Antognetti, P. and Milutinovic, V (1991) *Neural Networks - Concepts, Applications and Implementations,* Vol 4, Prentice Hall, Englewood Cliffs, N. J
9.   Aptronix (1993)   Fuzzy Logic from Concept to Implementation, Aptronix
10. Arbib, M. A(1995) *The handbook of Brain Theory and Neural Networks*, MIT Press, Cambridge, Mass
11. Banks, J., Carson, J. S. and Nelson, B. L(1996) *Discrete-Event System Simulation Second Edition*, Prentice Hall, Englewood Cliffs, N. Y
12. Barr, A (1981) The handbook of Artificial Intelligence 1, Addison-weslely, Mass
13. Barr, A (1982) The handbook of Artificial Intelligence 2, Addison-weslely, Mass
14. Barr, A (1089) The handbook of Artificial Intelligence 3, Addison-weslely, Mass
15. Barr, A (1989) The handbook of Artificial Intelligence 4, Addison-weslely, Mass
16. Bezdek, J. C. and Pal, S. K (1992) Fuzzy Models for Pattern Recognition : Methods that Search for Structures in Data, IEEE, N. Y
17. Bezdek, J. C(1991)   *Fuzzy Models for Recognition*, IEEE, N. Y
18. Bien's System Control Lab (1992) Topics on Fuzzy Logic-based Control and Applications 2, KAIST

19.   Bosc, P. and Kacprzyk, J(1995) *Fuzziness in Database Management Systems*, Physica-Verlag, Heidelberg

20.   Carpenter, G. A (1992) *Neural Networks for Vision and Imaging,* MIT press, Cambridge, Mess

21.   Cherkassky, V., Friedman, J. H. and Wechsler, H (1994)   *From Statistics to Neural Networks*, Springer-Verlag, Heidelberg

22.   Cho, J. W (1994) *Computer Architecture*, KAIST

23.   Cox, E (1994) The Fuzzy Systems Handbook - A Practitioner's Guide to Building, Using, and Maintaining Fuzzy Systems, AP Professtional, Boston

24.   David, R. and   Alla, H (1989) *Du Grafcet aux Rèseaux de Perti*, Hermes, Paris

25.   Dayhoff, J. E (1990) *Neural Network Architectures - An Introduction*, Van Nostrand Reinhold, New York, N. Y

26.   Dubois, D. and Prade, H (1980) *Fuzzy Sets and Systems : Theory and Applications*, Academic Press, New York

27.   Dubois, D., Prade, H. and Yager, R. R (1997) *Fuzzy Information Engineering : A Guided Tour of Applications*, John Wiley & Sons, New York

28.   Evans, G. W., Karwowski, W., and Wilhelm, M. R (1989) *Applications of Fuzzy Set Methodologies in Industrial Engineering*, Elsevier, Amstergam

29.   Fishwisk. P. A (1995) Simulation Modeling Design and Execution Building Digital Worlds, Prentice Hall, Englewood Cliffs

30.   Freeman, J. A. and Skapura, D. M (1991) Neural Networks algorithms, Applications and Programming Techniques, Addison-Wesley, Mess

31.   Gen, M. and Cheng, R (1997) *Genetic Algorithms and Engineering Design,* John Wiley & Sons, New York

32.   Ghezzi, C., Jazayeri, M. and Mandrioli, D (1991) *Fundamentals of Software Engineering*, Prentice-Halll, N. J

33.   Goldberg, D. E (1989) Genetic Algorithms in Search Optimization & Machine Learning, Addison-wesley, Mass

34.   Goonatilake, S. and Khebbal, S (1995) *Intelligent Hybrid Systems*, John Wiley & Sons, New York

35.   Gupta, M. M. and Yamakawa, T (1988) *Fuzzy Computing - Theory, Hardware, and Applications*, Elsevier Science Pub. Co., Amsterdam

36.   Gupta, M. M. and Yamakawa, T (1988) *Fuzzy Logic in Knowledge-based Systems, Decision and Control*, Elsevier Science Pub. Co., Amsterdam

37.   Gupta, M. M. et al (1985) *Approximate Reasoning in Expert Systems*, Elsevier Science Pub. Co., Amsterdam

38.   Gupta, M. M (1979) *Advances in Fuzzy Set Theory and Applications,* Elsevier Science Pub. Co., Amsterdam

39.   Gupta, M. M (1982) *Fuzzy Information and Decision Processes*, North-Holland, Amsterdam

40.   Gupta, M. M. and Sanchez, E (1982) *Fuzzy Information and Decision Processes*, North-Holland Pub. Co., Amsterdam

41.   Gupta, M. M., Saridis, G. N. and Gaines, B. R (1977) *Fuzzy Automata and Decision Processes*, Elsevier North-Holland, New York

42. Gupta, M. M. and Yamakawa, T (1988) *Fuzzy Computing Theory, Hardware, and Applications*, North-Holland, Amsterdam

43. Gupta, M. M. and Yamakawa, T (1988) *Fuzzy Logic in Knowledge-Based Systems, Decision and Control*, North-Holland, Amsterdam

44. Hajec, P (1998) *Metamathematics of Fuzzy Logic*, Kluwer Academic Publishers, Boston

45. Hertz, J (1991) Introduction to Theory of Neural Computation, Santafe Institute

46. Holland, J. H (1992) *Adaptiation in Natural and Artificial Systems*, MIT Press, Cambridge, Mass.

47. Horowitz, E. and Sahni, S (1978) *Fundamentals of Computer Algorithms*, Computer Science Press, Potomac, Md

48. Jain, A. K (1989) *Fundamentals of Digital Image Processing*, Prentice Hall, Englewood Cliffs, N.J

49. Jamshidi, M., Vadiee, N. and Ross, T. J (1993) *Fuzzy Logic and Control Software and Hardware Applications,* vol 2, Prentice Hall, Eaglewood Cliffs, N. J

50. Jamshidi, M., Vadiee, N. and Ross, T. J (1993) *Fuzzy Logic and Control Software and Hardware Applicartions*, PTR Prentice Hall, Englewood Cliffs, N. J

51. Jang, J. S. R (1997) *Neuro-fuzzy and Soft Computing*, Prentice Hall, Englewood Cliffs, N. J

52. Jang, J. S. R., Sun, C. T. and Mizutani, E (1997) Neuro-Fuzzy and Soft Computing : A Computational Approach to Learning and Machine Intelligence, Prentice-Hall, Upper Saddle River, N.J

53. Janko, W. H., Roubens, M. and Zimmermann, H. –J (1990) *Progress in Fuzzy Sets and Systems*, Kluwer Academic Publishers, Boston

54. Johes, A., Kaufmann, A. and Zimmermann, H. –J (1986) *Fuzzy Sets Theory and Applications*, D. Reidel Publishing Company, Dordrecht

55. Jung, J. Y., Im, Y. T. and Lee, K. H (1994) Fuzzy Approach to Shape Control in Cold Rolling of Steel Strip, *Electronics Letters*, Vol. 30, No. 21, 1807-1808

56. Jung, J. Y., Im, Y. T. and Lee, K. H (1996) Fuzzy Control Simulator of Cross Sectional Shape in 6-high Cold Rolling Mill, *Journal of Materials Processing Technology*, Vol. 62, 61-69

57. Jung, J. Y, Im, Y. T. and Lee, K. H (1995) Development of Fuzzy Control Algorithms for Shape Control in Cold Running, *International Journal of Meterials and Product Technology,* Vol. 48, 187-195

58. Kacprzyk, J. and Fedrizzi, M (1990) Multiperson Decision Making Models Using Fuzzy Sets and Possibility Theory, Kluwer Academic Publishers, Boston

59. Kandel, A (1986) *Fuzzy Mathematical Techniques with Applications*, Addison-wesley Pub. Co., Mass

60. Kandel, A (1994) *Fuzzy Control Systems,* Library of Congress Cataloging in publication data

61. Kandel, A (1991) *Fuzzy Expert Systems*, Crc Press, Boca Raton, FL

62. Kandel, A (1986) Fuzzy Mathematical Techniques with Applications, Addison-wesley Pub. Co., Mass

63. Kandel, A (1992) *Hybrid Architectures for Intelligent Systems*, Crc Press, Boca Raton, FL

64. Kandel, A (1986) *Fuzzy Mathematical Techniques with Applications*, Addison-wesley Pub. Co., Mass

65. Kandel, A (1986) *Fuzzy Techniques in Pattern Recognition*, John Wiley & Sons, New York

66. Kandel, A. and Lee, S. C (1979) *Fuzzy Switching and Automata : Theory and Applications*, Crane, Russak & Company, New York

67. Kaufmann, A., Dubois, T. and Cools, M (1975) *Exercises Avec Solutions sur la Théorie des Sous-ensembles Flous*, Masson et C, Editeurs, Paris

68. Kaufmann, A. and Gupta, M. M (1988) *Fuzzy Mathmatical Models in Engineering and Management Science*, North-Holland, Amsterdam

69. Kaufmann, A. and Gupta, M. M (1985) *Introduction to Fuzzy Arithmetic Theory and Applications*, Van Nostrand Reinhold Company, New York

70. Kaufmann, A (1977) *Introduction a la Théorie Des Sous-Ensembles Flous Tome 1*, Masson et C, Editeurs, Paris

71. Kaufmann, A (1975) *Introduction a la Theérie Des Sous-Ensembles Flous Tome 2*, Masson et C, Editeurs, Paris

72. Kaufmann, *A (1975), Introduction a la Théorie Des Sous-Ensembles Flous Tome 3*, Masson et C, Editeurs, Paris

73. Kaufmann, A (1977) Introduction a la Théorie Des Sous-Ensembles Flous Tome 4, Masson et C, Editeurs, Paris

74. Kaufmann, A., Zadeh, L. A. and Swanson, D. L (1975) Introduction to the Theory of Fuzzy Subsets, Vol 1, Academic Press, New York

75. Khanna, T (1990) Foundations of Neural Networks, Addison-weslely, Mass

76. Kim, C. B., Seong, K. A. and Lee, K. H (1998) Design and Implementation of Fuzzy Elevator Group Control System, IEEE Transactions on Systems, Man and Cybernetics, Vol. 28, No. 3, 277-287

77. Kim, C. B., Seong, K. A., Kim, J. O., Lim, Y. B. and Lee, K. H (1995) A Fuzzy Approach to Elevator Group Control System, IEEE Trans. on Systems, Man and Cybernetics, Vol. 25, No. 6, 985-990

78. Kim, J. K., Cho, C. H. and Lee, K. H (1998) A Note on the Set-theoretical defuzzification, Fuzzy Sets and Systems, Vol. 98, 337-341

79. Kim, J. K., Lee, K. H. and Yoo, S. W (2001) Fuzzy Bin Packing Problem, Fuzzy Sets and Systems, Vol. 120, No. 3, June, 429-434

80. Kim, Y. D. and Lee, K. H (1997) High Speed Flexible Fuzzy Hardware for Fuzzy Information Processing, *IEEE Trans. on Systems, Man and Cybernetics,* Vol. 27, No. 1, 45-56

81. Kim, Y. D., Park, K. H. and Lee, K. H (1995) Parallel Fuzzy Information Processing System, *Fuzzy Sets and Systems*, Vol. 72, 323-329

82. Kinnear, K. E (1994) *Advances in Genetic Programming,* MIT Press, Mass

83. Klir, G. J. and Folger, T. A (1988) *Fuzzy Sets, Uncertainty and Information*, Prentice Hall, Englewood Cliffs

84. Kohonen, T (1992) Adaptive Vector Quantization and Neural Networks, IEEE, Piscataway, N. J

85. Kosko, B (1997) Fuzzy Engineering, Prentice Hall, Upper Saddle River, N. J

86. Kosko, B (1993) Fuzzy Thinking, Hyperion, New York, N. Y

87. Kosko, B (1992) *Neural Networks and Fuzzy Systems*, Prentice Hall, Englewood Cliffs, N.J

88. Kosslyn, S. M. et al (1992) *Frontiers in Cognitive Neuroscience*, MIT Press, Cambridge, Mass

89. Koza, J. R (1992) *Genetic Programming*, MIT Press, Cambridge, Mass

90. Koza, J. R (1994) *Genetic Programming 2*, MIT Press, Cambridge, Mass

91. Langten, C. G (1995) *Artificial Life - an Overview*, MIT Press, Cambridge, Mass

92. Langten, C. G (1989) *Artificial life 1*, MIT Press, Cambridge, Mass

93. Langten, C. G (1992) *Artificial life 2*, MIT Press, Cambridge, Mass

94. Langten, C. G (1994) Artificial life 3, MITPress, Cambridge, Mass

95. Lee, C. C (1990) Fuzzy Logic in Control Systems : Fuzzy Logic Controller – Part I , *IEEE Trans. on Systems, Man and Cybernetics*, Vol. 20, No. 2, 404-418

96. Lee, C. C (1990) Fuzzy Logic in Control Systems : Fuzzy Logic Controller – Part II , *IEEE Trans. on Systems, Man and Cybernetics*, Vol. 20, No. 2, 419-435

97. Lee, J. H. and Lee, K. H (1999) Distributed and Cooperative Fuzzy Controller for Traffic Intersection Group, Part C, Applications and Reviews, *IEEE Trans. on Systems, Man and Cybernetics*, Vol. 29, No. 2, May, 263-271

98. Lee, J. H. and Lee, K. H (2001) Comparison of Fuzzy Values on a Continuous Domain, *Fuzzy Sets and Systems*, Vol. 118, No. 3, March, 419-428

99. Lee, K. H. and Lee, J. H (1999) A Method for Ranking Fuzzy Numbers and its Application to Decision-making, *IEEE Trans. on Fuzzy Systems*, Vol. 7, No. 6, December, 677-685

100. Lee, K. H. and Oh, G. R (1996) *Introduction to Systms Programming*, Prentice Hall, Singapore, London

101. Lee, K. H., Song, Y. S. and Lee, K. M (1994) Similarity Measure between Fuzzy Sets and between Elements, *Fuzzy Sets and Systems*, Vol. 62, 291-293

102. Lee, K. M. and Lee, K. H (1995) Identification of Lamda-fuzzy Measure by Genetic Algorithms, *Fuzzy Sets and Systems*, Vol. 75, 301-309

103. Lee, K. M., Kwak, D. H. and Lee, K. H (1995) Tuning of Fuzzy Models by Fuzzy Neural Networks, *Fuzzy Sets and Systems*, Vol. 76, 47-61

104. Lee, K. M., Kwak, D. H. and Lee, K. H (1996) Fuzzy Inference Neural Network for Fuzzy Model Tuning, *IEEE Trans. on Systems, Man and Cybernetics*, Vol. 26, No. 4, 637-645

105. Lee, K. M., Kwak, D. H. and Lee, K. H (1994) A Fuzzy Neural Network Model for Fuzzy Inference and Rule Tuning, *International Journal of Uncertain Fuzziness and Knowledge Based Systems*, Vol. 2, No. 3, 265-277

106. Leondes, C. T (1999) *Fuzzy Theory Systems Techniques and Applications*, Vol.3, Academic Press, San Diego

107. Li, D. et al (1990) *A Fuzzy Prolog Database System*, Wiley, Taunton

108. Liu, B (1999) *Uncertain Programming*, John Wiley & Sons, New York

109. Mamdani, E. H. and Gaines, B. R (1981) *Fuzzy Reasoning and its Applications*, Academic Press, London

110. Michalewicz, Z (1992) *Genetic Algorithms + Data Structures = Evolutionary Programs*, Springer-Verlag, Berlin

111. Michie, D. et al (1994) Machine Learning, Neural and Statistical Classification, Ellis Horwood limited

112. Miller, R. K (1987) *Neural Networks,* SEAI Technical Publications

113. Mitchell, M (1997) *An Introduction to Genetic Algorithms*, MIT Press, Cambridge, Mass

114. Miyamoto, S (1990) *Fuzzy Sets in Information Retrieval and Cluster Analysis,* Kluwer Academic Publishers, Boston

115. Moigene, J. L (1973) *Les Systèms D'information dans les Organisations*, Press Universitaires de France, Paris

116. Molloy, M. K (1989) *Fundamentals of Performance Modeling*, Macmillan Publishing Company, New York

117. Negoiţ ă , C. V(1975) *Application of Fuzzy Sets to Systems Analysis*, Wiley, New York

118. Negoiţ ă , C. V. and Ralescu, D. A (1975) *Applications of Fuzzy Sets to Systems Analysis*, Brikhauser Verlag, Basel

119. Nilsson, N. J (1980) *Artificial Intelligence*, Springer-Verlag ,Berlin,N. Y

120. Novak, V (1989) *Fuzzy Sets and Their Applications*, Adam Hilger, Bristol and Philadelphia

121. Park, S. W. and Lee, K. H (2000) Tape-2 Fuzzy Hypergraphs Using Type-2 Fuzzy Sets, *Journal of Advanced Computational Intelligence*, Vol. 4, No. 5, August, 362-367

122. Pedrycz, W (1989) *Fuzzy Control and Fuzzy Systems*, Research Studies Press, Taunton

123. Pedrycz, W (1993) *Fuzzy Control and Fuzzy Systems sec., Extended edition*, John Wiley & Sons, New York

124. Pedrycz, W (1995) *Fuzzy Sets Engineering*, CRC Press

125. Reisig, W., *Petri Nets*, Springer-Verlag, Berlin Heidelberg, 1985

126. Rich, E. and Knight, K (1991) *Artificial Intelligence Second Edition* McGraw Hill, New York

127. Ruan, D., *Fuzzy Logic Foundations and Industrial Applications*, Kluwer Academic Publishers, Boston, 1996

128. Rucker, R (1993) *Artificial Life Lab*, Waite Group Press

129. Ruspini, E (1992) *Introduction to Fuzzy Set Theory and Fuzzy Logic : Basic Concepts and Structures*

130. Schwefel, H. P (1995) *Evolution and Optimum Seeking*, John Wiley & Sons, New York

131. Scott, A. C. and Klahr, P (1992) *Innovative Applications of Artificial Intelligence 4*, MIT Press, Cambridge, Mass

132. Stowinski, R (1998) *Fuzzy Sets in Decision Analysis, Operations Research and Statistics,* Kluwer Academic Publishers, Boston

133. Sugeno, M (1985) *Industrial Applications of Fuzzy Control*, North-Holland, Amsterdam

134. Terano, T. et al (1987) *Fuzzy Systems Theory and Its Application,* Fajii Shisutemu Nyumon
135. Terzopoulos, D. et al (1997) *Artificial Life for Graphics, Animation, Multimedia, and Virtual Reality,* Siggraph 97
136. Tesauro, G. et al (1995) *Neural Information Processing Systems 7,* MIT press, Cambridge, Mass
137. Tesauro, G. et al (1996) *Neural Information Processing Systems 8,* MIT press, Cambridge, Mass
138. Walliser, B (1977) *Systèmes et Moedlès : Introduction Critique a L'analyse de Systèms,* Editions Du Seuil, Paris
139. Wang, L. X (1997) *A Course in Fuzzy Systems and Control,* Prentice Hall, Englewood Cliffs, N.J
140. Wang, L. X (1994) *Adaptive Fuzzy Systems and control: Design and Stability Analysis,* PTR Prentice Hall, Englewood Cliffs, N.J
141. Wang, P. P (1983) *Advances in Fuzzy Sets, Possibiliry Theory, and Applications,* Plenum Press, New York and London
142. Wang, P. P. and Chang, S. K (1980) *Fuzzy Sets : Theory and Applications to Policy Analysis and Information Systems,* Plenum Press, New York and London
143. Wasserman, P. D (1989) *Neural Computing - Theory and Practice,* Van Nostrand Reinhold, New York
144. West, D. B (1996) *Introduction to Graph Theory,* Prentice Hall, Upper Saddle River, N.J
145. Yager, R. R. and Zadeh, L. A (1994) *Fuzzy Sets, Neural Networks and Soft Computing,* Van Nostrand Reinhold, New York
146. Yager, R. R., Ovchinnikov, S., Tong, G. M. and Nguyen, H. T (1987) *Fuzzy Sets and Applicarions : Selectes Papers by L.A. Zadeh,* John Wiley & Sons, New York
147. Yang, N., Wohn, K., and Lee, K. H (1999) Modeling and Recognition of Hand Gesture Using Colored Petri Nets, Part A, Systems and Humans, *IEEE Trans. on Systems, Man and Cybernetics,* Vol. 29, No. 5, September, 514-521
148. Zadeh, L. A (1998) *The Life and Travels with the Father of Fuzzy Logic,* TSI Press
149. Zadeh, L. A.and Kacprzyk, J (1992) *Fuzzy Logic for the Management of Uncertainty,* John Wiley & Sons, New York
150. Zimmermann, H. –J (1984) *Fussy Sets and Decision Analysis,* Elsevier Science Publishers, New York
151. Zimmermann, H. J (1985) *Fuzzy set Theory and Its Applications,* Kluwer-Nijhoff Publishing, Boston
152. Zimmermann, H. J (1991) *Fuzzy Set Theory and Its Applications : Second Revised Edition,* Kluwer Academic Publishers, Boston
153. Zimmermann, H. J (1987) *Fuzzy Sets and Decision Making and Expert Systems,* Kluwer Academic Publishers, Boston
154. Zimmermann, H. J., Zadeh, L. A. and Gaines, B. R (1984) *Fuzzy Sets and Decision Anaysis,* North-Halland, Amsterdam
155. Zurada, J. M (1994) *Introduction to Artificial Neural Systems,* West-Publishing, St. Paul

Index

Printing and Binding: Strauss GmbH, Mörlenbach